*Altering Houses*

*and*

*Small-scale Residential Development*

# ALTERING HOUSES
## and
# Small-scale Residential Development

*Ann and Colin Bridger*

OXFORD   BOSTON   JOHANNESBURG   MELBOURNE   NEW DELHI   SINGAPORE

Butterworth-Heinemann
Linacre House, Jordan Hill, Oxford OX2 8DP
225 Wildwood Avenue, Woburn, MA 01801-2041
A division of Reed Educational and Professional Publishing Ltd

A member of the Reed Elsevier plc group

First published 1998

**British Library Cataloguing in Publication Data**
Bridger, Ann
    Altering houses and small-scale residential development
    1. Dwellings – Maintenance and repair   2. Architecture, Domestic
    I. Title   II. Bridger, Colin
    690.8

ISBN 0 7506 4100 2
Printed and bound in Great Britain by
Biddles Ltd, Guildford and King's Lynn

# Contents

# Preface

Nearly everyone is interested in domestic property. At the least it forms the background for daily life and is often the focus of aspirations. The working life of thousands of people is centred in this field. Houses embody ideals as well as space for activities. They also absorb a great deal of the country's wealth.

Thus the ups and downs of the housing market affect people financially, and their images of comfort and security, of leisure and worthwhile work are in the balance. The need for shelter and security, an acquisitive instinct and the urge to create a distinct and personal environment all tend to propel people into buying and altering houses.

When the domestic property market is dormant there may seem less immediate impetus for activity but the underlying reasons for interest are as strong as ever. However, because there is no guarantee that prices rise with inflation in the short term, or even that employment patterns and government policy will favour owner occupation as in the recent past, it is necessary to act with caution and be very clear about the objectives for development and methods of building.

The idea for this book came in response to a request for a general guide to buying a house to live in and for information on how to profit from updating and altering the property. Design, technical and commercial factors, which are usually studied separately, needed to be considered together. Also, the legislative constraints on development and the costs of building works and associated transactions had to be addressed. Hence the aim of this book is to provide clear and straightforward guidelines to the complex business of small-scale residential development, i.e. buying and altering houses.

This book is primarily intended for people who wish to manage a project themselves and who may even wish to become physically involved. However, few people would have the time or skill needed to undertake all the complex processes and there are many cases where professional assistance is essential, so the input of the various professions is also considered. It is anticipated that it will be a useful general text for students of the various disciplines and others who want to know how the different processes of small-scale development interrelate.

The usual relationship between these complex activities has suggested a structure for the book; however, the way in which it is used will vary with individual knowledge and needs. For ease of reference each chapter is divided into sections and each section is divided into parts.

Throughout the book an emphasis has been placed on timing and costing because the development process can be very slow, yet the fluctuations in the property market often occur suddenly. A study of the condition of buildings and common defects is also an important element in the text.

Although the book is mainly about altering houses, many of the considerations are equally applicable to new work and it is hoped that the text will be useful for people working on new projects.

This book aims to cover a wide field in a concise way, only adding some background information to set matters in context, and therefore does not deal in great depth with many of the issues involved; however, extra reading is recommended at the end of each chapter.

We should like to acknowledge the help of many people who have given their time and advice in the preparation of this book. We should also like to thank Don Fisher, Teresa Hill, Ruth Lay and Richard Purkiss for their help in producing the text and drawings.

**Ann and Colin Bridger**

# Appraisal and design

This chapter deals with the important factors to be considered at the outset of a residential development project. It begins with a discussion on the purpose of undertaking such a project and explains the factors which influence the selection of a suitable property for improvement or extension. The most important sources of information related to a property are reviewed and guidelines are provided on how to approach a detailed site investigation. Finally an outline of the most relevant design procedures is provided. These are the procedures by which the project moves from being feasible to being buildable. Consultants may become involved and what they seek to achieve is discussed finally.

## 1.1 Establishing a purpose for development

Development suggests a wide range of activities, many on a very large scale, but in this book development means those activities in which the form of a dwelling is altered to achieve an objective. The objectives vary and can be incorporated with works of repair and adaptation. Since many large extensions are virtually new buildings, many of the principles can be applied to this form of development as well.

Bearing in mind that building owners have altered buildings for as long as there have been buildings to alter, we shall consider in this section only the reasons which currently appear to influence owners to undertake what are often rather complex operations.

### 1.1.1 Pleasure

Many people derive a great deal of pleasure from working or having work done on their homes, particularly if the property is upgraded or increases in value as a result. The frustrations and inconvenience often experienced during the course of the works are clearly overcome by the satisfaction felt with the result. A 1997 report (Davidson, Redshaw and Mooney (1977) *The Role of DIY in Maintaining Owner-occupied Stock*. Policy Press, University of Bristol) showed that in 1991 some £4 billion was spent by owner-occupiers on materials and tools for decorating, repairing and upgrading their dwellings. This is a major contributor to the UK economy. Apart from DIY enthusiasm there is much enjoyment to be experienced from seeking out properties 'in need of improvement', investigating potential sites and weighing the advantages and disadvantages of particular localities. Clearly people do not always move for the most practical of reasons.

## 1.1.2 The need to 'personalize' an environment

Many people have neither the resources, the opportunity nor the desire to move house to achieve satisfaction with their domestic surroundings. If by effecting an improvement an urge to personalize their homes can be satisfied then the benefits are considerable. Small schemes of this kind include:

- adjusting the experience of daylight: more with larger windows, less with smaller windows, considerably more with a conservatory extension;
- adapting a kitchen to suit a lifestyle: minimalist approach or country style, welsh dressers and quarry tiles; cooking to be enjoyed; an abundance of appliances (to remove the drudgery); experimenting with new techniques;
- adapting the bathroom: making it decorative; fittings of a style to suit the house; bath or power showers or both; bigger baths; the trend to 'en suite' arrangements;
- changing spaces: open plan living spaces or separate rooms;
- redecoration, often unnecessary but refreshing, inexpensive and immensely popular.

## 1.1.3 Spatial needs

Domestic accommodation rarely wholly satisfies the spatial needs of the occupier. People's requirements vary enormously and few have the luxury of a 'made to measure' home. Changes in lifestyle can make some rooms too small, in the wrong place or in the wrong relationship. Typical of such changes are:

- more people to accommodate: a new baby, an elderly relative, an au pair perhaps;
- an office is needed: someone needs to work at home using a computer network, fax machines and photocopiers, for example;
- entertaining of friends or business colleagues perhaps: more space needed in the 'reception' rooms;
- health and fitness becoming popular: space in the basement possibly for a jacuzzi, aerobics, table tennis and so on;
- independence: youngsters needing separate rooms at an earlier age than they would have in the past.

## 1.1.4 Upgrading

This is the most popular reason for undertaking DIY work and forms a part of most alteration and extension jobs. When the prevailing economic conditions deter householders from moving house, they will tend to make the best of what they have and there is a great deal of guidance available on what can be done and how to do it. Popular forms of upgrading include:

- improving thermal insulation, particularly by laying down insulating materials in the roof space and double glazing the windows;
- providing an en suite bathroom or shower room to the main bedroom in addition perhaps to a 'family' bathroom;
- providing a ground floor cloakroom with a shower;
- installing an energy-efficient heating system – using a condensing boiler perhaps and thermostatic controls;
- adding a conservatory as a link between house and garden and, if facing south, making some use of solar energy;
- adding a garage or car port, at the same time incorporating storage for an increasing use of bicycles;
- improving the appearance of the property – new materials to replace old, new paintwork, new fences, putting everything in good repair and so upholding its value;
- replacing obsolescent and potentially defective services such as electrical wiring, fittings, pipework and so on but at the

same time increasing the capacity to accommodate modern appliances and accessories;

- replacing kitchen fittings, introducing new appliances and creating a dining kitchen;
- converting the loft for more than just storage;
- providing access for disabled people;
- improving security by the introduction of security locks, lighting, alarm systems and so on.

## 1.1.5 *Income or capital appreciation*

The prospect of an alteration or extension providing additional income may add further weight to the householder's desire to effect some improvements to his or her home. Contrary to popular belief, improvements do not always increase the value of a property in financial terms and may even decrease it if done badly. As we shall see in the next section, there are many factors affecting value. Making alterations pay is a more tangible benefit. Some of the ways that can be used are listed below.

### *Income*

- Income from letting a room or rooms to a lodger; taxes change, of course, but in early 1998 up to £4250 per annum could be obtained tax free in this way.
- Income from holiday lets; clearly in appropriate locations only and generally through an agency which will require a minimum standard of accommodation to be provided.
- Providing bed and breakfast; standards are improving such as en suite bathrooms, television in all rooms and so on; hard work for the hosts, a degree of uncertainty but seasonal.
- An assured shorthold tenancy; for a small apartment perhaps, where separate facilities and access are available; a degree of security.
- As a continuous activity; buying, improving and selling properties in succession; the role of a developer.

### *Capital appreciation*

- Buying a house, living in it and selling. An attractive option when the housing market is buoyant. This usually involves no Capital Gains Tax liability.

- Buying a house, letting it and selling. An alternative option, particularly where the rent obtained can be set against costs and the sale can be delayed until the economic conditions are favourable.

## 1.2 Selecting a property

Three of the several factors which affect the selection of a property for development are worth particular attention. They are:

- Location
- Neighbours
- Potential or latent value

In this section we discuss their importance.

### 1.2.1 Location

It is often said that only three things matter in buying a property – location, location and location. This is because a location embodies the image of a lifestyle and it is that image which is the most fundamental, but often the most difficult to define, reason for selecting a property. Improvements in transport and communications may have affected some of the practical considerations for location but the importance of access to a wide range of facilities which affect lifestyle would appear to be growing. People have been known to approach an estate agent with a list of factors they will not accept, such as semi-detached or half-timbered, but end up buying a house with just those features because the location and image of a particular house were right.

Householders will appreciate that whereas they can change their properties, within certain legal constraints, they can do very little to change their surroundings, so looking for the location that best fits their requirements in the first place makes good sense. The low value of an area as a whole may reduce the value of any single property, however much is spent on it. On the other hand, the value of a location can be improved by changes in its infrastructure as we shall see in Part 1.2.3.

The location factors which most affect selection fall into two main categories which can be summarized as follows.

**Practical factors**

- Public transport availability – bus, rail, air etc.
- Accessibility to major road networks
- Availability of shopping centres
- Availability of schools and other education facilities
- Availability of leisure facilities – sport, theatre, cinema, boats etc.
- Accessibility to public open spaces for dog walking and so on

**Environmental factors**

These are the rather indefinable things which affect the quality of an area such as:

- Spaciousness or density of development
- Quality of infrastructure
- Relationship between buildings and landscape
- Noise, particularly road traffic and aircraft
- Exposure to extreme weather conditions
- Air quality and risk of pollution generally

## 1.2.2 Neighbours

Disputes with neighbours over boundaries, trees and so on are one of the most common sources of litigation. A careful inspection of the boundaries of a property is a wise precaution. Potential problems such as overhanging trees or dilapidated sheds and fences should be noted. Investigations in due course will establish whether neighbouring owners enjoy any particular rights affecting the property, such as shared access, rights of way or drainage. Talking to the people concerned should be an opportunity for any grievances to be aired and agreement to be reached on possible future difficulties.

Detecting some problems with neighbours can be difficult. These are the ones which can only be discovered by other neighbours 'informing' on them or by actually occupying the property for some time, preferably over the summer months.

Typical of these are the following.

### Noise

- Loud voices, shouting, arguments – poorly built terraced housing of any age can be surprisingly deficient in sound insulation.
- Music – the type seems to matter less than the volume and the propensity of those who enjoy loud music to play it with open windows. Outside, the garden wall does little to help.
- Children – playing noisily and out of control can be a source of aggravation to some people.
- DIY activity – persistent use of an electric saw or a hammer drill can be troublesome, particularly if it occurs at unsociable hours and if the activity is concerned with the neighbour's livelihood.

### Smells

- An enthusiasm for barbecues, strongly spiced food, fish, deep frying and so on can blind people to the effect they can have on a neighbour's enjoyment of fresh air, particularly where extractor fans are badly directed and the external space is rather confined. Cooking smells may fall somewhat short of air pollution but if persistent they can nevertheless cause annoyance.

The condition of a neighbouring property may give cause for alarm. Peeling paintwork, broken windows, blocked guttering and other forms of dilapidation can have a depressing effect on the value of houses nearby. The problem is often a lack of or poor maintenance and since this is usually the responsibility of the owner (who may not be the occupier) it is advisable to enquire who owns the property and to ascertain the reasons for the apparent lack of care. As a general rule owner-occupiers are the most likely to keep their houses in good repair but there are always exceptions to the rule.

In these comments on neighbours we have assumed that the neighbourhood is a residential one. Clearly in the situation where the neighbouring property is non-residential and in particular is an active industrial or commercial site, the problems can be much worse. Night working can cause great annoyance and should be investigated immediately.

Persistent noise offenders can be dealt with by taking action in the following order:

1 Talk to the person or company responsible. If that fails:
2 Consult the local mediation service if there is one. Information on this can be obtained from Mediation UK, 82a Gloucester Road, Bishopston, Bristol BS7 8BN, tel: 0117 924 1234. If that fails:
3 Complain to the local authority Environmental Health Department which has considerable powers, including the serving of an abatement notice and taking the offender to court.
4 As a last resort and if the local authority does not take action, it is possible under Section 82 of the Environmental Protection Act 1990 for an owner to complain directly to a magistrates' court. The environmental health officer will still offer advice in this case. A court may impose a fine of up to £5000 in a domestic case or £20 000 if industrial, trade or business.

### 1.2.3 Potential or latent value

There is still a useful profit to be made from buying a property with latent value and waiting or acting to realize that value. Latent value can be affected by changes in the locality, improved facilities or accessibility, for example. The rehabilitation of derelict land and buildings, the opening or closing of schools and pedestrianization schemes may all have an effect. It can also be altered in quite subtle ways, by changes in people's perception of an area or by the social standing of the residents perhaps.

Oxford is an interesting example. At one time, Jericho, a neighbourhood within a mile to the north east of the city, was the area in which to look for convenient low priced housing with a social cachet because of its proximity to the colleges and its North Oxford ambience. Property values rose in spite of a density of development many would describe as claustrophobic. Grandpont, a somewhat blighted area of similar size on the south side of the city, in contrast benefited from innumerable local changes, especially the landscaping of derelict land close to the River Thames and a number of riverside developments. As a result Grandpont became as desirable as Jericho as a place to live, albeit for different reasons.

### Latent value opportunities

- A run-down building in a 'rising' area, i.e. an area where the potential value has not yet been widely appreciated.
- A house on a plot large enough to permit another dwelling or large extension to be built (subject to planning permission).
- A large house suitable for conversion to two or more units (subject to local planning policy being favourable).
- A house with superficial problems which are not difficult to resolve, e.g. garish paintwork.
- A house which needs simple repairs, e.g. faulty renders or pointing.
- A house which has been poorly improved in the past, e.g. steel windows in an old cottage.
- A house which is underpriced 'for a quick sale', an executor sale perhaps.
- A house which is poorly presented, e.g. surrounded by clutter and dilapidated outbuildings.
- A house which needs landscaping, e.g. new boundary walls, an improved driveway or more appropriate planting.
- A house in a location about to benefit from a change of infrastructure, e.g. a new motorway link.
- Any building which provides an opportunity to create an imaginative and attractive dwelling; mills, barns and redundant workshops, for example.

## 1.3  Sources of information

Having located a house with potential for development it is necessary then to obtain as much information about it and its locality as possible. Should the purchase go ahead in due course then other enquiries will be made in the conveyancing process, as we shall see in Chapter 8. The main sources of information in the first place are likely to be estate agents, local planning authorities, taxation and environmental authorities, branches of county libraries, tourist offices, neighbours and local residents.

The gas, electricity, water and telecom companies will provide information on services, particularly those which are concealed below the ground with little or no evidence on the surface of their existence.

### 1.3.1 Estate agents

The asking price for a property on the market is usually provided but in some circumstances such as a rapidly rising or falling market a guide price might be given instead. Where there are similar houses in the vicinity it is possible to check if the range is right but, as we have seen, the condition of the property may make a difference to its value. Vendors usually want as high a price as possible, except in the case of an executor's sale in which case the beneficiaries are generally in a hurry to ensure that the property does not stand vacant. Agents try to guide clients to a realistic figure but in the end may err either way, mainly because the property market fluctuates with supply and demand and other influences such as the availability of money and changing interest rates.

The estate agent's particulars should state whether a property is freehold or leasehold, the council tax band, whether it is a listed building or in a conservation area and the size of the plot. However, some or all of this information may be omitted, in which case questions must be asked. If the property is leasehold, the ground rent and the annual maintenance charge should be known. It is important to know the overheads at the outset otherwise time and effort could be wasted if they prove prohibitive.

The type of property, i.e. detached, semi-detached, terraced, its location, approximate age, use in the recent past, the size of the rear garden and the number and overall dimensions of the main rooms are usually given in the particulars. The width of the hall, any corridors and the dimensions of bathrooms and small kitchens are often omitted even though they may be critical when planning alteration work. General information on the amenities of the locality is usually given in broad terms, such as 'convenient for shops'; useful guidance but not totally reliable.

It is a legal requirement that estate agents' particulars should be accurate. It has been known for an agent to be taken to court for describing a garden as south facing when it was not.

However, the legislation has had the effect of reducing the amount of information agents are now prepared to give and has resulted in even more caveats such as 'all dimensions are approximate', or 'we have not tested any equipment'; thus the presence of a central heating boiler is no indication of its condition or ability to function.

## 1.3.2 Local authority offices

### Planning

A visit to the offices of the Planning Department of the Local District Council will determine whether:

- the property is in a conservation area
- the building is 'listed'
- any trees on or adjacent to the site are the subject of a Tree Preservation Order
- an Article 4 Direction is applicable
- there have been previous applications to develop the property.

These factors will undoubtedly affect proposed development in ways which are fully discussed in Chapter 5.

The local planning authority keeps records of past and current applications for properties in the area and it is possible to inspect these for guidance as to what has been approved or refused locally and consequently what might be acceptable or unacceptable in the future. At the same time it may help to clarify whether any existing extensions to dwellings could have been done as 'permitted development' in which case no application for permission was necessary.

Local, district or town plans should be available for inspection at local authority offices. These will reveal the authority's policies with regard to new housing schemes in particular areas, or its policies on conservation and so on. Local plans are frequently updated so it is important to take note of recent revisions.

### Building control

The Building Control department of the local authority keeps records of past applications and approvals under the Building Regulations which it should be possible to view as a prospective purchaser of a particular property. This is an aid in identifying parts of a building which have been altered or extended and a rough guide as to the construction and materials used. It should be remembered, however, that original construction drawings are not always an accurate record of the 'as built' condition.

Meeting the area building control officer can be helpful, particularly with respect to any particular local conditions such as 'made ground', land liable to flood, industrial pollution, methane or radon gas and so on.

### Highways

Public highways are maintained by the County Council Highways Department which must be consulted if alterations are to be made to any public road or, in particular, if a new access is intended. When an application for planning approval is submitted to the District Council and this includes a change of access to a road, the Highways Department of the County Council is automatically consulted so there is no need to consult them directly. On the other hand the Highways Department will readily provide information on the design and construction of roads, footpaths and crossings which meet with its approval and this can be helpful at the survey stage.

### Council tax

The local District Council is responsible for administering and collecting council tax although a large proportion of the tax is passed on to the County Council to cover such services as education, social services, highways and the fire service. To determine the tax liability for any particular property one should contact the relevant department of the District Council, probably now called Revenue Services.

### Sewerage

Information on local sewers (main drains usually in the highway) can be obtained from the local water company from whom permission must be obtained before making any connection to a public sewer. Permission to install a septic tank or cesspool (see Part 2.5.5) must be obtained from the relevant regional office of the Environment Agency which will provide full information on its requirements.

### 1.3.3 *Local residents*

People who live or work nearby will probably know the advantages and disadvantages of the neighbourhood, local schools and shops, transport and parking facilities, local nuisances and so on. They will know of seasonal changes to the local environment which will not be apparent to a first-time visitor. Long-standing residents may also know of any incidence of flooding, severe weather damage, ground movements and other hazards which may have some bearing on the proposed development. Apart from that, local history is often very interesting if one is proposing to live in the neighbourhood.

## 1.4 Site appraisal and investigation

If the early investigations into the feasibility of a project seem favourable it will be necessary to look more closely at the site, to determine what opportunities it has to offer and to appreciate its constraints. Building sites vary enormously: on the one hand the 'open field' type, on the other the small back garden, with many variations in between. The principles of site investigation are much the same in any case but our emphasis here is on the rather small and restricted piece of land on which an extension or one or two houses are to be built. Where the work is confined to the interior of a house a site investigation may not be necessary except where foundations are affected, as we shall see later. Smallness is not a reason for overlooking a site's characteristics but it does reduce the importance of some of the factors considered.

Where consultants are involved with the design of a project, an early site investigation is essential. An architect or surveyor can make little progress in design work without fully understanding the effect that site conditions may have on the design. Even in speculative house building where standard house plans are used, the foundations cannot be completely standardized because sites offer such variable ground conditions.

The activities that comprise a complete site investigation fall into three phases: desk studies, a preliminary site appraisal and a full site investigation. These can be described as follows.

## 1.4.1 Desk studies

These are an extension of the earlier enquiries described in Section 1.3. Once again the aim is to collect as much information as possible before going to the site. Time spent at this stage is well spent because it can inform the later stages of the investigation and in some cases serious difficulties can be exposed early enough for the project to be abandoned with minimum expenditure or loss of time.

Desk studies can include:

- *a study of old maps* – early Ordnance Survey maps often reveal features such as buildings, water courses, woodland and made ground (or fill), all of which may no longer be apparent on the surface but can affect ground conditions and therefore foundation design and construction.
- *a study of geological maps* – the information to be gathered from these often requires interpretation by a geologist but an indication of what subsoil types are likely to be found in the site area can be helpful. Spotting the presence of shrinkable clays, alluvial deposits (in river valleys) or peat-bearing soils, for example, can not only affect the nature of a subsequent soil survey but can give some indication to a contractor that work below ground may not be straightforward. The 1 inch to 1 mile geological maps can be used for these early studies. These are available in the central libraries of most towns, as are the Ordnance Survey maps mentioned above. Anyone who wishes to learn more about local geology should consult the British Geological Survey which organization publishes a great deal of detailed information for most areas and has an excellent library at the Geological Museum in London.
- *a study of aerial photographs* – these can be useful in detecting past building works and in archaeology they have a particular purpose in this respect. Where past photographs are available and the dates are known they may illustrate the

history of a site, particularly with regard to trees that have been felled, buildings or extensions to buildings that have been demolished and other changes of a topographical nature.

## 1.4.2 Preliminary site appraisal

To obtain accurate and current site information a visit is essential. The desk studies carried out beforehand will have highlighted some of the important areas to be investigated but many details can only be picked up by keen observation on the site itself. One problem is that the preliminary site appraisal must be undertaken quickly, probably in one day, particularly if the site is some distance from the investigator's office. It is therefore essential to approach the task systematically so that nothing escapes notice which may have a bearing on the activities which follow on. It is after the first site inspection that an architect, for example, will begin design work, even though a more thorough investigation may occur in due course. A systematic approach will also ensure that a record is made of all observations, which is very important where the information is to be used by people who have not visited the site themselves. For these reasons a checklist approach is considered useful and some appropriate headings for this are suggested below.

In the first place, and to avoid the necessity of a second visit, it is important to go to the site suitably equipped. A professional surveyor will employ a selection from the following:

- Boots
- A camera/camcorder
- A voice recorder and/or clipboard and paper
- A spade and assorted tools for probing the surface
- A manhole cover lifter
- A sound level meter
- An automatic level and staff
- Measuring tapes and rods
- A compass
- A short ladder
- Small containers to carry away samples of soil, water etc.
- Binoculars

A recommended method is to begin with a fairly 'broad brush' approach taking in impressions of the whole site and work down gradually to the detail.

### The checklist with associated notes

1 *Location of site* – check that it corresponds with the one on the map or other particulars. Mistakes have been made.

2 *Boundaries of site* – check that they conform with maps and so on. Note their existence, construction, height and condition. Beware of unsafe garden walls.

3 *Existing trees* – check against Tree Preservation Order (if any), then for all trees their approximate age, species, condition and proximity to proposed or existing buildings. Assume that a tree is a plant which has a stem exceeding 75 mm in girth.

4 *Existing buildings* – identify those known to be 'listed', then note the approximate age, type, size, position, condition and use of all buildings on the site.

5 *Adjacent properties* – as for 4 as far as possible but in addition note their proximity to the boundary, any apparent party walls and any structures which would appear to be at risk from nearby construction work. Also note the position of windows in neighbouring properties potentially enjoying rights of light or creating perhaps an overlooking problem.

6 *Orientation and aspect* – the OS map will show the orientation of the site but note the effect that trees and the height and form of buildings could have on the amount of sunlight available, taking account of summer and winter conditions. Note potentially interesting views from different levels and positions on the site and any aspects of the site likely to be exposed to abnormal weather conditions. The prevailing winds in most of the UK are south westerly but local variations can occur.

7 *Site levels* – changes of level or slopes should be recorded, particularly if they are likely to affect the development. Simple assessments of rise and fall can be made with a spirit level and batten or more accurately an automatic level and staff. Steeply sloping sites are best left to a surveyor using land surveying equipment.

8 *Services* – note any superficial evidence of services, which should correspond with information received from the

supply companies. This would apply to gas, electricity, water, drainage, cable TV and telecommunications. Underground services will be difficult to detect except where inspection chambers have been provided. Overhead lines, however, present a hazard, perhaps quite early on when an excavator arrives to dig a trial pit. Also, they and the poles that support them may have to be relocated, at the developer's considerable expense. If any gas, electricity or water services appear to be live on a derelict or vacant site, the supply companies should be informed immediately.

9   *Obstructions below ground* – a superficial inspection of the ground may reveal traces of past buildings. When buildings are demolished, their foundations are usually left in place and if there had been a cellar this will be filled (or part filled and therefore dangerous) with rubble of doubtful quality. Building over such sites requires the advice of a structural engineer who may suggest special foundations. Unfortunately these things do not always appear on maps or old records and are not detected until building work has started. Old wells, air raid shelters and ice houses are uncovered from time to time and if considered of historic importance they will attract the attention of the local archaeological society and the conservation officer of the local authority, which may cause an unexpected delay pending further investigations.

10  *Soil conditions* – in advance of a full soils survey, which is discussed later, there are only superficial observations which can be made on the first visit to site. Note any evidence, for example cracking, on buildings or walls nearby, showing that settlement has been taking place and note the type and condition of vegetation on the site. Lush green grass and reeds close to a river might suggest a high water table, for example. Digging out a spit or two of soil may provide more clues, particularly about top soil, but is no substitute for a deeper investigation.

11  *Access* – note the existing points of access, both pedestrian and vehicular, and assess the possibility of their reuse. An awareness of County Highway standards will help in deciding on the suitability of an existing access on to a public road. The width of a gateway, space for turning vehicles and

sight lines are typical considerations not only for the long-term use of the site but also for safety in the use of contractors' vehicles and delivery lorries.

12  *Local facilities or amenities* – these have been mentioned earlier but a site visit may provide an opportunity to confirm the existence of local schools, sports facilities, public transport and so on. Travelling to the site to carry out the survey can give some indication as to its accessibility but one must be mindful of the fact that conditions can vary with the day of the week and the time of the year.

13  *General environmental quality* – this is even more inclined to vary with the time of the visit but being on the site on a normal working day should enable a person with reasonably perceptive powers to detect persistently high noise levels or poor air quality and on a more positive note to record the 'green' features, the birdsong and other intangible qualities which make up the 'atmosphere' of an area.

Figure 1.1 shows part of a record of a preliminary site appraisal.

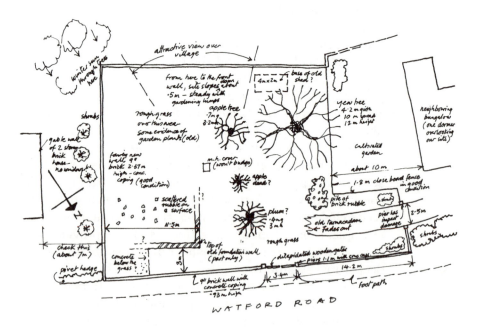

*Figure 1.1   A preliminary site appraisal*

### 1.4.3 The full site investigation

For small low-budget schemes the preliminary site appraisal we have described will generally be sufficient. A small domestic extension, for example, may have little impact on its site, or vice versa, but the importance of an adequate bearing for foundations should not be overlooked, however small the project. For larger schemes and those in which adverse site conditions are anticipated, a more searching investigation may be required. For this specialists are needed and the cost of employing them can be justified where the value of the job overall is sufficiently high. The objective in site investigations is to reduce the risk of things going wrong later in the project. The combination of low value and high risks often presents the client with a difficulty – is the relatively high cost of further investigation justified?

A full site investigation usually comprises the following:

1   *A further study* of the most relevant factors from the preliminary stage, summoning assistance where necessary. For example:

   ● looking at suspect trees in the company of the local authority forestry officer;
   ● re-measuring noise levels in the company of the local authority environmental health officer;
   ● inspecting neighbouring properties, particularly party walls, in the company of the adjoining owners or their surveyors;
   ● pegging out the footprint of a proposed building or extension so that the relationship of building and site can be more fully appreciated. Even the new building's height can be indicated by the use of a long pole;
   ● re-checking the condition of services in the company of representatives from the supply companies.

2   *A building survey* – a detailed study of a building and its immediate environs often undertaken by a professional building surveyor and comprising:

   ● a measured survey – a record of what exists is generally useful, but essential in the case of a historic building;
   ● a condition survey – the general condition of the building fabric and services;

- a structural appraisal – an inspection of the building's structure, its stability, its movement behaviour and the likely effect of alterations – undertaken jointly with a structural engineer;
- a valuation – a requirement where a property is being bought and sold.

These activities are more fully discussed in Chapter 2.

3  A *site survey* – a measured survey of a piece of land either an open field or restricted in close proximity to buildings. This will indicate:

- site levels and, on large sites, contours;
- surface features, buildings, trees, shrubs and so on;
- boundaries and all elements set out to a precise scale.

The charge for the survey of a typical house plot is likely to be several hundred pounds, depending of course on its complexity.

4  A *soils investigation* – very much the most sophisticated form of surveying at this level and therefore the most expensive. A full soils survey for a small housing scheme may cost several thousand pounds. The investigation is carried out by a specialist firm of engineers who produce a report of their findings. This takes time but, as we have seen, the preliminary investigation undertaken earlier should provide sufficient pointers for design work to proceed in the meantime. Ground conditions and their influence on foundation design are more fully discussed in Section 2.7.

## 1.5  Aspects of design

### 1.5.1  Design objectives

Building alterations, however small, always include an element of design and the way in which this is carried out depends on several factors. The amount and complexity of the work, the time available and the level of funding will all affect what is required and who should do it. Above all, it is the client's perception of the importance of design which will dictate whether the process is anything more than the production of a simple drawing.

The client may have very clear ideas and even produce sketches of what is wanted but usually the services of someone with design communication skills will be required to convert those sketches into workable schemes. Certainly if the client's ideas are uncertain then professional help will be necessary and a brief must be established. Architects are well used to such situations and will see it as an exchange of ideas – the client stating in broad terms what the 'user requirements' are while the designer, with due regard to budget and site constraints, suggests ways in which the requirements can be satisfied. In this way both client and designer can begin to work on preliminary design ideas together and begin to understand each other's roles.

There are many objectives to be achieved in the design of a building or an alteration scheme, the most practical of which is the production of drawings. The number of drawings required for a job is generally in proportion to its value, so that for a small domestic extension only one may suffice but for a single new house several will be needed. The number is affected by the scale of the drawing and the size of the sheets and both may be affected by easy access to copying facilities. Whatever the number to be produced, it is important to realize that drawings are produced for different purposes as shown in Table 1.1.

Table 1.1 shows that design drawings are usually developed in stages. The preliminary design may comprise only a few sketches for the purpose of gaining the client's approval; they may not show every facet of the design. The scheme design is a later development, having been worked out in detail and then presented using colour and other effects in order to win the approval of planning committees and the like. Design drawings do not normally show constructional detail or services unless the scheme is small enough for all necessary information to be contained on one sheet. See Figure 1.2.

Production or working drawings can be produced in any order but it is convenient to divide them so that some show the plans, sections and elevations of both the existing and proposed building (usually to scales of 1:50 or 1:100) and others show the components and how they are assembled (usually to scales of 1:5, 1:10 or 1:20). See Figure 1.3. Ideally no working drawing should be considered complete until others relating to it have

21

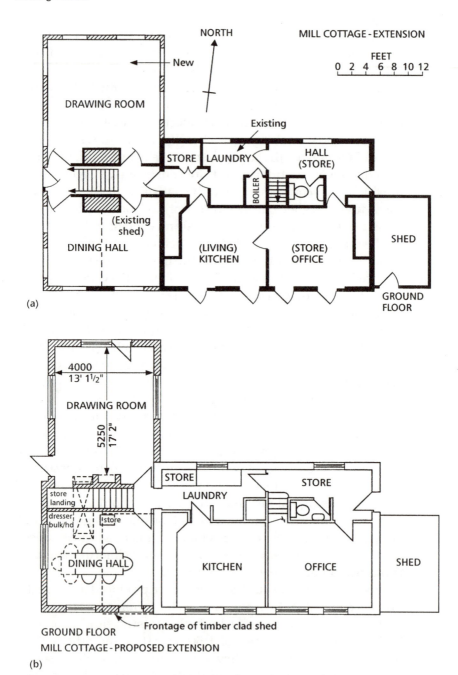

**MILL COTTAGE - EXTENSION**

NORTH

New

DRAWING ROOM

Existing

STORE | LAUNDRY | HALL (STORE)

BOILER

(Existing shed)

DINING HALL

(LIVING) KITCHEN

(STORE) OFFICE

SHED

GROUND FLOOR

FEET
0 2 4 6 8 10 12

(a)

4000
13' 1½"

DRAWING ROOM

5250
17' 2"

store
landing

dresser
bulk/hd

store

DINING HALL

STORE

LAUNDRY

STORE

KITCHEN

OFFICE

SHED

GROUND FLOOR

Frontage of timber clad shed

MILL COTTAGE - PROPOSED EXTENSION

(b)

*Figure 1.2 Design drawings: the client's sketch (a) is the basis of the designer's preliminary design (b)*

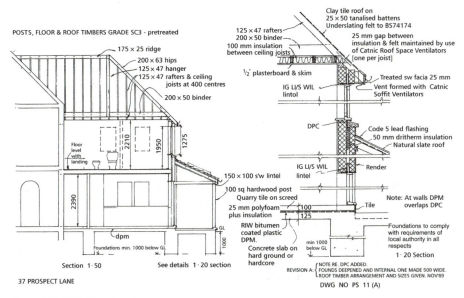

POSTS, FLOOR & ROOF TIMBERS GRADE SC3 - pretreated

175 × 25 ridge
200 × 63 hips
125 × 47 hanger
125 × 47 rafters & ceiling joists at 400 centres
200 × 50 binder

Floor level with landing

2210
1950
1275

2390

dpm
Foundations min. 1000 below GL

Section 1·50    See details 1·20 section

37 PROSPECT LANE

Clay tile roof on
25 × 50 tanalised battens
Underslating felt to BS74174

125 × 47 rafters
200 × 50 binder
100 mm insulation between ceiling joists

25 mm gap between insulation & felt maintained by use of Catnic Roof Space Ventilators [one per joist]

½″ plasterboard & skim

IG LI/S WIL lintol

Treated sw facia 25 mm
Vent formed with Catnic Soffit Ventilators

DPC

Code 5 lead flashing
50 mm dritherm insulation
Natural slate roof

IG LI/S WIL lintel

Render

150 × 100 s'w lintel
100 sq hardwood post
Quarry tile on screed
25 mm polyfoam plus insulation
RIW bitumen coated plastic DPM.
Concrete slab on hard ground or hardcore

GL
1000
100
125
min 1000 below GL
Tile

Note: At walls DPM overlaps DPC

Foundations to comply with requirements of local authority in all respects

1·20 Section

NOTE RE. DPC ADDED.
REVISION A: FOUNDS DEEPENED AND INTERNAL ONE MADE 500 WIDE.
ROOF TIMBER ARRANGEMENT AND SIZES GIVEN. NOV'89

DWG NO PS 11 (A)

*Figure 1.3   Working drawings*

## Table 1.1   *Drawing types and purposes*

| Stage | Drawings | Typical purposes | Producers |
|---|---|---|---|
| Design | Sketch or preliminary design | Client's approval; informal approval of local planning authority; cost planning; early project planning | Architect, surveyor, technician |
| | Full scheme design | Further client's approval; full planning permission; approval of funding bodies | As above |
| Production | Basic production or working drawings | Building regulation approval; cost checking; obtaining quotations from sub-contractors and suppliers; design coordination; contract documents | Architect, surveyor, technician, sub-contractors, mechanical and electrical engineers, structural engineer |
| | Full and detailed working drawings | Measurement for bills of quantities; estimating and tendering; site construction; contract documents | As above |
| Use | 'As built' drawings adapted from those above | Health and safety file (owner's manual) as required by CDM regulations | As above |

at least been 'roughed out'. Kitchen walls, for example, should not be 'fixed' until the kitchen fittings have been detailed. In fact, kitchen fittings and other manufactured items such as windows, doors, sanitary ware and ironmongery are best described with the aid of schedules which are useful for the purposes of pricing and ordering.

For a new building project or sizeable extension, drawings alone will not provide all the information required. For tendering or construction purposes, statements about materials and workmanship are necessary so these should be incorporated in a 'specification'. For example, a drawing will show the location and dimensions of a concrete floor but the specification will describe the mix and quality of the concrete to be used in its construction. Specifications are discussed again in Chapter 9.

### 1.5.2  Design procedures

So far, emphasis has been placed on the aspect of design which is concerned with the production of information for a number of different purposes. Information in the form of drawings, schedules and specifications will normally be produced by a 'design team' in which an architect, surveyor, engineer and sub-contractors may be involved. Their work must be systematic and well co-ordinated to be effective. The membership of the team and the person who leads it will be influenced by the following considerations:

- The client's own expertise as a designer; it has been known for an architect to employ another architect but most will opt for the economy of DIY.
- The importance attached to design by the client; one who is very particular may require high standards and be prepared to pay for them.
- The size and complexity of the scheme – the larger or more complex, the more necessary it becomes to appoint designers.
- Evidence or experience of satisfactory previous arrangements, for example a contractor/designer relationship that has worked well before.
- The advice given by the leading designer or the first professional to be consulted (not necessarily the same person).

- The availability of consultants when needed, the level of fees they charge and their 'conditions of engagement'.

Since the mid-1960s the Royal Institute of British Architects (RIBA) has published guidelines on job management, including a 'Plan of Work'. This is essentially a framework for the architectural service and fee structure but it also serves as a guide on design procedures generally. See Appendix A.

Inevitably it emphasizes the architect's role but it is important to realize that other professional roles may be just as or more important to the client. In particular, a client who sees costs as a priority may appoint a quantity surveyor as a design leader; one seeking a quicker result may approach a contractor in the first place.

Figures 1.4 to 1.6 illustrate basic design procedures where the overall objectives are similar but the priorities are somewhat different in each case. The contract is assumed to be a house extension costing in the order of £30 000 and involving new heating and electrical installations. Figure 1.4 shows typical procedures where an architect is appointed as the design team leader. Characteristics of this are:

- The architectural design is of paramount importance.
- The architect is in the traditional role of managing, controlling and co-ordinating the whole process. He or she is therefore ideally suited to act as a 'planning supervisor' under the Construction Design and Management (CDM) regulations.

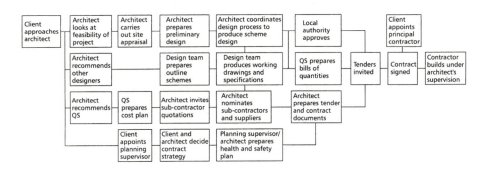

*Figure 1.4   Design procedures: architect appointed*

- The contractor joins the process late and therefore has no input to design and can only negotiate on price, not design.
- The architect being wholly responsible for design may enter into a contract with the client to that effect.
- The architect is responsible for site supervision.
- The client can depend on the design team to consider feasibility, develop a brief and produce alternative proposals and costings.
- Competitive tendering is involved.
- The architect nominates sub-contractors and oversees their design work.

Figure 1.5 shows an alternative procedure whereby a quantity surveyor is appointed as design team leader. This has the following broad characteristics:

- Strict cost planning and control are essential to the purpose of the project.
- The quantity surveyor manages the whole process, a high priority being given to the economics of the scheme. He or she may act as a CDM planning supervisor.
- The architect may be the key designer but no longer manages or controls all procedures.
- The quantity surveyor (acting as a 'contract administrator' in contract terms) may be responsible for such matters as the nomination of sub-contractors and dealing with variations.
- The contractor is once again late in joining the process.
- The quantity surveyor is responsible for site supervision.

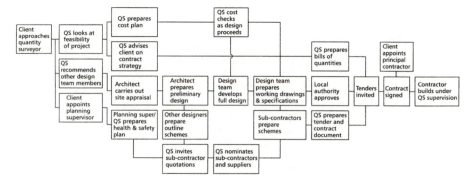

*Figure 1.5   Design procedures: quantity surveyor appointed*

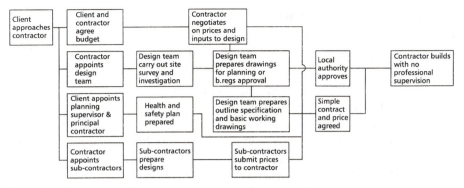

*Figure 1.6  Design procedures: contractor appointed*

Figure 1.6 shows a relatively simple case where the client has entrusted the design process to a contractor offering a design service (a 'design and build' approach). Particular features of this are:

- The design is relatively straightforward – simple form and specification, repetitive work and so on.
- The contractor manages the whole process and decides what design skills are needed.
- The contractor is responsible to the client for the design, including that provided by sub-contractors.
- The designers are low profile, producing drawings and outline specifications only. Detailed specifications may be a matter of chance.
- The designers have no responsibility for site operations.
- The client is totally reliant on the contractor for design, quality and cost control unless they appoint a consultant to represent them in such matters.
- The client should approach this process with a precise brief or at least a clear idea of what is needed.
- The price for the job is 'negotiated', not the subject of competitive tendering.

As we shall see in later chapters, the variations and complexities of these procedures are not evident from these simple diagrams. Nevertheless it is helpful to be constantly aware of overall design objectives and the broad roles of those involved in achieving them, particularly when the mind becomes preoccupied with complications later.

### 1.5.3 Design principles

To suggest that the design process is only about producing information would be to ignore the creative aspects of design which are important even in the smallest schemes. Attitudes to building design can vary enormously; on the one hand, some will value the element of mystery which surrounds design and which is encouraged to some extent by the professions. On the other hand, others see it as an essentially practical process of satisfying functional requirements. Plain and simple building design is probably somewhere in between, requiring as it does a good deal of imagination but also a good knowledge of construction and a respect for the environment.

For the purpose of this book, a systematic and practical approach is considered the most useful. Design flair can never be ruled out, rather it is to be encouraged, but not at the expense of functional considerations. Neither can questions of cost, practicality and safety be ignored at any stage of the process. A summary of the most important functions in building design is given in Table 1.2. It would be interesting to analyse each of these in some detail but that is outside the scope of this book. Here only brief explanatory notes are given for the most important of them.

#### Contractual

- The designer's contract with the client – agreeing the level and quality of design information required.
- The contract between the client and contractor – deciding the extent to which the contractor and sub-contractor are to be responsible for design.
- Assessing the capabilities of the contractor and sub-contractors – of interpreting design information and executing work of the quality required.

#### Technical

- Selecting the combination of structural form and materials which can best provide a building fit for its purpose within an acceptable cost.
- Buildability – designing a building which can be constructed efficiently, safely, economically and to required levels of quality.

### Table 1.2    The functions of building design

*Design functions*

| | |
|---|---|
| Contractual | Client/designer |
| | Client/contractor |
| | Contractor's capabilities |
| Technical | Structure/materials |
| | Buildability |
| | Site constraints |
| | Legislation |
| | Environmental impact |
| Financial | Size, form and specification |
| | Costs-in-use |
| | Energy efficiency |
| Functional | Geometric/spatial |
| (user requirements) | Structural |
| | Environmental |
| | Subjective:    Aesthetics (scale, form, proportion, symbolism) |
| | Comfort |
| | Privacy |
| | Protection/security |
| | Control/territory |
| | Self-expression |
| | Well-being |

- Site constraints – taking account of slopes, soils, water, trees, access and so on.
- Legislation – designing within the constraints of planning, building regulations, CDM regulations, party walls and so on.
- Environmental impact – designing a building which is environment friendly.

## Financial

- Taking account of the relationship between design and cost with particular regard to the effects on cost of size, form and specification.
- Giving due consideration to costs-in-use (running costs) as opposed to initial costs (capital outlay) and how they relate to maintenance and servicing.
- Energy efficiency – designing a building which uses materials with low 'embodied energy' and is not wasteful in the use of energy.

## Functional

- Considering the users of the building (not necessarily the owners), including visitors and those who clean and maintain – their need for a building that works well.

- Analysing user requirements – all the characteristics of a building which support its function such as:

  - *Geometric/spatial*: adequate space of suitable form or shape with good access and circulation spaces. Economic designs attempt to eliminate superfluous space without adversely affecting comfort or aesthetic appeal.
  - *Structural*: a structural form of suitable material which will safely sustain and transmit loads to the ground (the building must stand up). Where appropriate the structure should be resistant to wear and tear, to impact loads and to misuse.
  - *Environmental*: adequate but not excessive levels of heating, lighting, ventilation, water and waste disposal. Minimum requirements are specified in the Building Regulations but there is much scope for 'green' design where insulation, efficient heating systems and careful use of energy are becoming increasingly necessary.
  - *Subjective*: the less tangible factors such as aesthetics or visual quality in terms of style, scale, form, proportion, colours and textures; the degree of comfort required, privacy, security and the means by which self-expression or a sense of well-being can be achieved. Sometimes so subtle that they lack the attention they deserve, these factors are nevertheless important to the user and should be considered in even the smallest of alteration schemes.

# Further reading

Baden-Powell, C. (1997) *Architect's Pocket Book*. Arch Press.

British Standards Institute (1997) *BS1192 – Part I: Construction Drawing Practice*. BSI, 1984.

Building Research Establishment (1987) *Site Investigation for low-rise building, Desk Studies (BRE Digest 318)*. HMSO.

Building Research Establishment (1997) *Climate and site development (BRE Digest 350)*. HMSO.

Cecil, R. (1993) *A Client's Guide to Building*. Legal Studies Publishing Ltd.

Department of the Environment (1997) *Bothered by Noise? There's no need to suffer*. HMSO.

Thompson, A. (1990) *Architectural Design Procedures*. Arnold.

Tutt, P. and Adler, D. (1997) *New Metric Handbook*. Arch Press.

# 2

# The survey

This chapter outlines different approaches to the detailed survey of an existing dwelling. Because elderly buildings are rarely fault-free, it goes on to describe some common building defects and remedies and some of the characteristics of older services installations. Then three of the most serious problems encountered in surveys are discussed in detail – rising damp, poor foundations and timber decay. Finally, some notes on listed buildings are added.

There are many different forms of surveying using very different techniques to achieve different objectives. Common to all, though, is the need for a keen sense of observation and accuracy of measurement. In general building work there are three types of surveying which are directly relevant:

- *Quantity surveying* – this is measuring the amount of work involved in a job, usually before it starts, by studying drawings and specifications and producing lists of items (bills of quantities) for pricing by contractors. The role of the quantity surveyor and its broader implications are discussed in Chapters 4 and 9.
- *Land surveying* – this is the measurement of land or the spaces surrounding buildings, using sophisticated equipment such as a theodolite, and producing drawings on which ground levels and surface features such as buildings and trees are shown. New building projects will almost certainly require a site survey to be carried out and specialist firms are available to do them. When the site is relatively simple, though, an architect, surveyor or technician in the design team will be able to produce a site plan which is sufficiently accurate.

- *Building surveying* – this is the measurement of a building both inside and out. Unlike the others, though, the building surveyor is frequently required to record the condition of the building and judgement is required as to the relative severity of any defects observed. The full role of the building surveyor is discussed in Chapter 4. It should be said, though, that some of the work of the building surveyor can be done by architects, contractors, engineers, technicians and owners themselves. Unlike land surveying, building surveying demands general competence rather than specialist skills and for this reason much of this chapter is devoted to it.

Building and land surveying are similar in one important respect. Both are concerned with producing drawings which show physical form as it exists, for the main purpose of designing modifications to that form. Accuracy of information is therefore very important. There is also the need on occasions to produce survey information purely as a record of what exists, particularly where historic buildings and sites are concerned, as we shall see later.

## 2.1 Measured building surveys

Most alteration and conversion jobs and even some repairs require an accurate survey drawing showing the plans, elevations and sections of an existing building. These will form the basis of design work and are frequently required as part of a planning or building regulation application. The size and type of building will dictate the approach and method used but some general guidance is given below.

### Recommended equipment

Folding rule; lockable measuring tape (5 m); spirit level; plumb bob (12 m terylene line); torch; clipboard + paper (5 mm squared is useful) or notebook; camera; overalls; steps or short ladder; telescopic height measuring rod; straight rod or batten; umbrella.

### Approach/method

This must be systematic and the information recorded in such a way that another person can pick it up and fully understand it. Skill and care are needed and getting it right takes time. Suggested procedure:

- Devise a method of identifying each room or space to be measured and its relationship to the whole. Points of the compass, or reference to front, back, left and right as viewed from the main approach perhaps. Also, of course, note the floor level. A typical room would be 'first floor front left' and so on.
- Stand in each room or space and draw a plan of it, getting the proportions as accurate as possible without actually measuring, although 'pacing out' the main dimensions and using squared paper will help (say 10 squares to the metre, which on 5 mm squared paper would be a scale of 1:20). If the rooms are small, several will fit on to one sheet but leave space for notes and dimensions. Mark on the plan the dimensions you wish to take and sketch any details to a larger scale (a door architrave, for example). Include diagonal dimensions corner to corner which help to get the shape right when the plan is finally drawn out. Remember that rooms that seem square are often not. Curved walls will need dimensions to be taken as offsets from a straight line. See Figure 2.1. Wall thicknesses can be measured at windows or door openings, floor thicknesses at landings and ceiling heights at the centre of rooms.
- Enlist the help of one or two people in taking measurements. Ideally have two people on the tape, one writing down the dimensions. Use 'running dimensions' which are quicker and more accurate than 'plain dimensions'. See Figure 2.2.

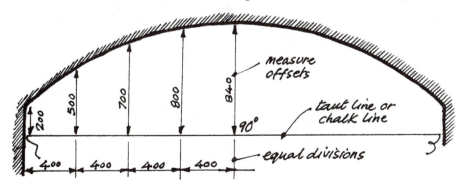

Figure 2.1  Measuring a curved wall by use of offsets

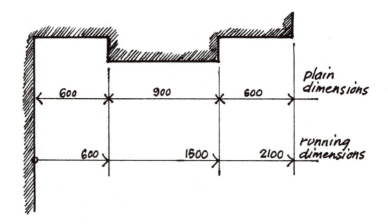

*Figure 2.2   Two ways of dimensioning an object*

Fixtures such as shelving and cupboards will need to be added. Externally, heights are sometimes a problem although a modern telescopic rod will measure up to 8 metres with ease. Counting courses is one way of dealing with the height of brick buildings. Measure the height of four typical courses and use this to calculate the height of the whole, e.g. 4 courses = 300 mm therefore 48 courses = 12 × 300 = 3.6 m. When measuring external heights, changes in ground level must be noted and measured. This can be done roughly by using a batten and spirit level but greater accuracy requires the use of an optical level. If a long ladder is necessary it should be used at the correct angle and someone should be available to 'foot' it. See Figure 2.3.

- Draw out the plans on A3 or A4 sheets (which are easy to photocopy) starting with the ground floor, to an appropriate scale, say 1:50 or 1:20 for a dwelling, 1:100 for something larger. Details can be drawn at 1:10 or 1:5. Once the ground floor has taken shape, the other floors should 'fit' over it but beware of upper walls which are unexpectedly displaced from lower walls. Building structures are not always as logical as they seem at first sight.
- Use the plans as an aid in drawing the elevations. If the measurements taken externally do not seem to correspond to the plans, the error should be traced even to the extent of remeasuring on site. It is normal practice to

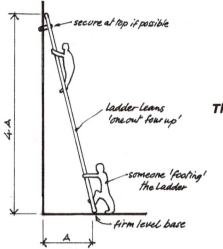

Figure 2.3   Correct use of a ladder (see also Figure 10.15)

label the elevations by reference to the points of the compass which implies, of course, that the plans are marked with a 'north point'.

## The results

Figure 2.4 shows a surveyor's typical site notes for part of the interior of a Victorian house. The following should be noted:

- Dimensions are plain rather than running.
- Height dimensions are encircled. In the case of windows H = head, C = cill.
- Use of diagonals as a check on 'squareness'.
- Details well related to main plan.
- Use of section to record heights and location of section shown on plan.
- Use of brief notes on some of the features.

## Problems

Surveying buildings is rarely plain sailing. Some of the problems encountered are as follows:

- Remembering to have all the equipment available so that only one site visit should be necessary.
- Deciding on the degree of accuracy. In this respect the survey should reflect the needs of the intended works. A kitchen should be measured accurately – to, say, the nearest 3 mm – because manufactured components must fit in. A roof space can be measured less accurately – to, say, the nearest 12 mm which is a reflection of the kind of tolerances a carpenter working on the roof can be expected to work to.
- Access. Occupied buildings are sometimes difficult to survey when furniture, personal belongings, vegetation and so on can slow the work down and lead to inaccuracies. Measured

35

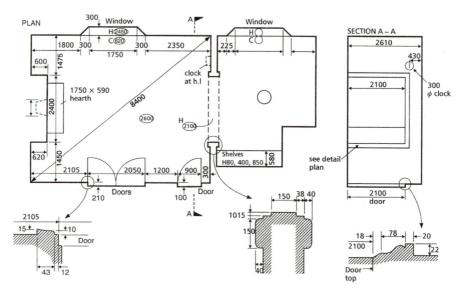

Figure 2.4 *Measured survey notes*

surveys do not require probing below the surfaces of things but the lack of an access hatch into a roof space can be frustrating.

- Taking care. A professional surveyor will be aware of his or her 'duty of care' which is more fully discussed in Chapter 4. In measured surveys the risk of physically damaging a property or its contents should be quite low but the potential damage caused by an error of measurement must be considered. Drawings are sometimes marked 'all dimensions to be verified on site' which means that a contractor must check the dimensions of a stairwell, for instance, before making a staircase. This disclaimer, though, cannot relieve the surveyor of his or her obligation to exercise 'reasonable' care.

## 2.2 Condition surveys

For the purpose of this book, a survey in which the main objective is to record the physical and environmental condition of a building is referred to as a condition survey. The reader should be aware, however, that the terms 'building surveying' and 'structural surveying' are often used to describe the same activity.

## Recommended equipment

This is much the same as for measured surveys with the following additions: manhole cover lifter; airtight containers for samples if necessary; binoculars for inspecting roofs and chimneys; voice or video recorder for more extensive note taking.

## Approach/method

The condition survey requires a considerable amount of note taking and it is useful to have available drawings of a scale which allows the notes to be related to particular elements on the plans. A preliminary measured survey would have made such drawings available and in practice the two surveys are often part of the same overall building survey. In any case a clear system of room identification should be used, as it was in the measured survey.

The survey should begin with a preliminary inspection of the whole property. This will help in deciding how to approach the work, its extent and possible difficulties. The building should be seen as a structural system, in which loads are transferred from roof to ground, and an environmental system in which the walls and roof contain a particular set of atmospheric conditions. One basic defect can lead to a flurry of others and only by looking at the whole building can the 'chain' of events be traced.

The next step is to go through the building room by room, making notes on the condition of the fabric as seen on each surface in turn – ceiling, floor, north wall, east wall and so on. Windows and doors may have to be numbered and should of course be opened and closed as a test of their function. In fact all mechanical devices, taps, WCs and so on should be tested in this way. The main objective is to locate and appraise defects as described in Section 2.4, but a comprehensive survey should also describe general condition in some way – as very good, good, fair, poor, for example. Unlike the measured survey, the condition survey should include an inspection of voids such as roof spaces, ducts and timber floors, the main purpose being to seek out any infestations of rot or beetles and to locate any weakening of the structure by careless work in the past.

Having completed the interior inspection, the same approach should be followed for the outside facades, dealing with them one by one. The condition of rain water goods should be noted, as should any evidence that they do not function well such as moss or algae growth. The roof can be inspected with binoculars but parts which are not visible such as behind chimneys or in valleys should be clearly noted. Faults such as defective leadwork or delaminating slates are almost impossible to detect unless access to the roof can be provided.

The survey may include an inspection of boundary walls and grounds and will almost certainly include the building's drains. In this case, unless a defect has been notified, it is usually sufficient for manhole covers to be lifted and one person to inspect them while another flushes a WC or two. If there appears to be something wrong then a specialist firm can be called in to test and report on the drains. See Part 2.5.5 on drainage.

### Results

Figure 2.5 shows an excerpt from a surveyor's site notes for part of the interior of a dwelling in rather poor condition and Figure 2.6 shows a similar entry for one facade of the building.

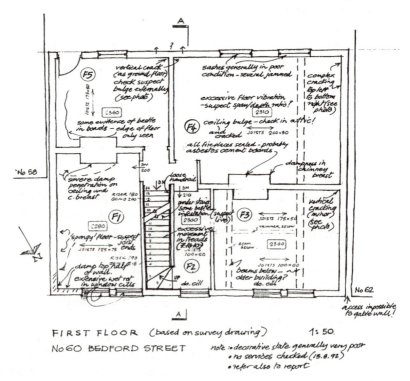

Figure 2.5   Surveyor's site notes: interior

### Problems

The problems experienced in measured surveys and described in Section 2.1 are common to most forms of surveying. Investigating the condition of a building may result in further complications such as:

- Inaccessibility of roofs and floors, particularly in occupied buildings, may result in the surveyor being unable to report on parts of the building and he or she will use a disclaimer clause in his or her report. General disclaimers are not popular so the surveyor should be precise in stating what has not been inspected, and that coupled with the exercising of 'reasonable care' is all that can be expected in this difficult situation.

- Surveying listed buildings requires the utmost care and the standard practice of probing surfaces with a penknife to ascertain their condition cannot be condoned. Neither is it possible to extract samples of materials for testing as one would in the normal course of events. The extent to which panels can be removed, floorboards lifted, holes dug and so

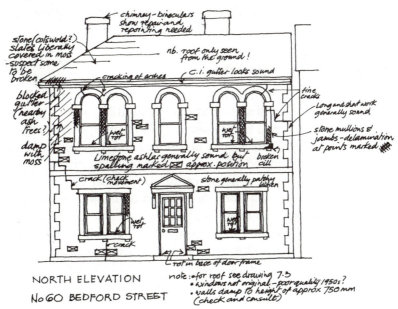

Figure 2.6    Surveyor's site notes: exterior

on should be discussed with the local conservation officer and a local archaeological society may well become involved as observers and consultants. It is the recording aspect of surveying which is important to conservationists and for this an understanding of historic buildings is considered necessary. See Section 2.9 for a further discussion on listed buildings.

- The diagnosis of faults in buildings can be difficult. It requires a keen sense of observation, good judgement and technical expertise. Experience and training are very important. Even so, the conditions observed may be so complex that specialist advice may be needed and the prudent surveyor will acknowledge the fact. Dangerous structures, heavy infestations of timber and dilapidated services are typical of the conditions where a second opinion would be advisable. Errors of judgement, oversights and miscalculations are all too frequent and in today's climate of litigation, most professional surveyors will protect themselves against claims of negligence by taking out professional indemnity policies. Regrettably, as is the case with other consultants, this has the effect of increasing overheads, and therefore charges to the client who may find the addition rather tiresome. We refer to this later in Section 4.4.

### The professional approach

As an example of a systematic approach to condition surveying, the 'Home Buyer Survey and Valuation' service promoted jointly by the Royal Institution of Chartered Surveyors (RICS) and the Incorporated Society of Valuers and Auctioneers (ISVA) is to be recommended. The service comprises a careful inspection of the property, a concise report based on the inspection and a valuation. Its main purpose is to assist prospective buyers of domestic properties and it is not intended to be used for unusual or dilapidated buildings which will require a more thorough and time-consuming survey. Nevertheless it is useful and in 1997 the RICS made the key elements of the service mandatory for its members. The essential framework of the Survey and Valuation report, as drafted for consultation in 1997, is given for guidance in Appendix B.

## 2.3 Structural appraisal

### Movements

Some building defects are of a structural nature. These are normally the result of movement in the building, the causes of which may be very unclear. The leaning wall shown in Figure 2.7, for example, may have been displaced by the outward thrust of weakened roof trusses, differential settlement of the soil below the foundation, extreme wind forces or, in all probability, a combination of all three. All buildings move from time to time, adjusting their shapes as a response to environmental changes. Buildings built before the middle of the nineteenth century, when the introduction of portland cement led to the use of rigid mortars and concrete, tend to be more flexible than their modern counterparts and movement should be expected and allowed for. The problem is deciding when movement becomes so pronounced as to become potentially damaging to the structure. The settlement of foundations is often a defect in this category.

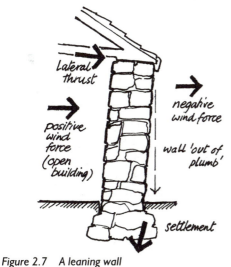

This is where the advice of a structural engineer who understands old buildings can be very useful indeed.

Another cause of structural movement can be the manner in which a building has been altered in the recent past. Buildings will adjust themselves to accommodate new stresses induced by additional or redistributed loads, but provided the alterations are carried out with care this should be a short-lived effect.

Figure 2.7   A leaning wall

### Cracking

Structural movement usually gives rise to cracking, particularly where brittle materials such as cement-rich mortars and renders are concerned. The direction and width of cracks can inform a

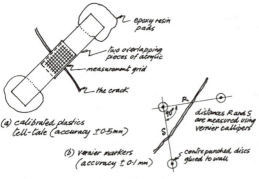

(a) calibrated plastics tell-tale (accuracy ±0.5mm)

(b) vernier markers (accuracy ±0.1 mm)

*Figure 2.8   Two methods of measuring movements in cracks*

trained eye of the direction in which the movement is taking place. Hair-line cracks are usually harmless but where any crack, even when apparently harmless, is allowing water to penetrate the structure then it should be sealed with an appropriate material (lime mortar in lime mortar brickwork, for example). One important question is whether cracks are 'dead' (movement has ceased) or 'live' (movement is continuing). If in doubt, a surveyor or engineer may use crack monitoring devices such as those shown in Figure 2.8. These would be in place for a period of time so that the direction and size of movement can be measured. Where it is suspected that the movement is seasonal, as is the case with a clay settlement problem, the monitoring should take place for at least a year so as to cover all seasons.

### The role of the structural engineer

As we saw in condition surveys, the need to appraise the building as a whole is essential in structural surveys. Discovering how a building works structurally, how the loads are transferred and where the stresses are concen-trated is not only useful but also very interesting. In an old build-ing it usually takes time to do this, more than would be necessary for an equivalent new building, but it is worthwhile nevertheless. Calculations are of doubtful benefit, although the local building control officer may ask for evidence that a particular element, a beam or column perhaps, is capable of carrying a proposed new load. A wise structural engineer will produce rudimentary calculations without too much reference to codes of practice which have little relevance to old structures. Where alterations are intended which include the use of proprie-tary components such as steel lintels or purlins, then the manufacturers of these will provide calculations when required.

When a structural engineer becomes involved in a survey it is likely that he or she will look at the structure in two ways:

- Does the existing structure need repairs?
- Do the proposed alterations appear to be structurally satisfactory?

After an initial survey the engineer may recommend:

- A documentary search – particularly for old drawings etc.
- Immediate emergency measures – avoiding a potential collapse
- Further investigation by opening up concealed spaces – as we have already discussed
- In-situ testing or materials testing
- Digging trial holes (see Section 2.7)
- Consulting specialists with regard to timber infestation and moisture control

## 2.4   A matrix of common faults and remedies

The following pages provide only a guide as to the likely causes and possible remedies for a series of common faults in small to medium-sized traditional buildings. It is important to realize that in practice fault diagnosis can be complex and each one should be appraised in situ and remedies suggested which are compatible with the structure and environment of the host building. Buildings are consistent in only one respect – they differ from each other.

For convenience the matrix is divided into four sections, Roofs, Walls, Floors and Foundations, but it will be apparent that there is considerable overlap between the sections, as one would expect. The remedial work suggested assumes that people with appropriate skills are available, but where workmanship or expertise is particularly important it is mentioned.

| Common faults | Likely causes | Remedies |
|---|---|---|
| *Roofs* | | |
| Broken tiles or slates | • Falling tree branches<br>• Walking on roof without protection<br>• Frost action if slates/tiles old | • Replace them quickly but safe practice demands that appropriate equipment is used |
| Leaking valley gutters | • Old lead, zinc or felt corroded | • Short-term patching<br>• Total removal in longer term<br>• Plumber's work |

| Common faults | Likely causes | Remedies |
|---|---|---|
| Leaking rain water pipes and gutters | • Blockages<br>• Fractures from frost, movement or impact | • Clear blockages regularly<br>• Replace broken sections<br>• Check falls of gutters |
| Leaking flat roof (common in post-1945 extensions) | • Ageing finish (e.g. felt)<br>• Impact damage<br>• Poor workmanship<br>• Inadequate falls | • Short-term patching<br>• Strip off and replace<br>• Check fall and rebuild decking if necessary |
| Cracked cement mortar listings (fillets at base of stacks) | • Shrinkage of cement mortar<br>• Structural movement | • Replace with gauged or hydraulic lime mortar or lead flashings (plumber's work) |
| Lead flashings separate from wall | • Thermal movement or deterioration of backing | • Pin back into joints using lead wedges and repoint |
| Loss of pointing in chimney stack | • Ageing and weathering<br>• Frost attack<br>• Sulphates in old chimneys (see Figure 2.9) | • Repoint with gauged or hydraulic lime mortar (not cement mortar) by skilled bricklayer |
| Spalling of bricks/stones in chimney stack | • Frost attack and/or salts<br>• Impervious cement pointing leads to excessive moisture penetrations<br>• Structural movement | • Cut out and replace with bricks/stones to match in colour, texture and permeability<br>• Mortar as for repointing (above) |
| Leaning chimney stack | • Ageing<br>• Wind forces<br>• Sulphates from flue gases (see Figure 2.9)<br>• Support failure | • Structural engineer or surveyor to assess risk of collapse<br>• Line flue, repoint or rebuild on firm base |
| Roof spreading outwards | • Decay or removal of timber tie beams or collars<br>• Overloading | • Structural engineer or surveyor to advise<br>• Timber repairs by competent carpenter<br>• Deal with any timber infestation |
| Roof sagging | • General decay of timber framework, particularly wall plates and ends of rafters/trusses<br>• Overloading | • As above<br>• Temporary propping may be needed<br>• Use treated timber |
| Condensation in roof space | • Warm moist air ascending from below and trapped in space insufficiently ventilated | • Improve ventilation<br>• Deal with source of moist air (within reason)<br>• Moisture check on ceiling below insulation<br>• Lag cold pipes and tanks (not below the latter) |
| Rotting thatch | • Ageing – long straw lasts, say, 20 years, water reed, say, 50 years<br>• Birds and squirrels can help decay | • Depends on extent of decay<br>• Can be repaired or total replacement plus protective netting<br>• Consult local master thatcher only |

*Figure 2.9   Sulphate attack on chimneys*

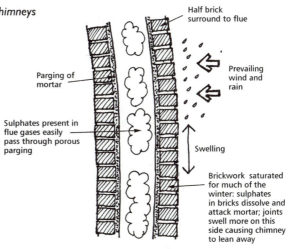

Half brick
surround to flue

Parging of
mortar

Prevailing
wind and
rain

Sulphates present in
flue gases easily
pass through porous
parging

Swelling

Brickwork saturated
for much of the
winter: sulphates
in bricks dissolve and
attack mortar; joints
swell more on this
side causing chimney
to lean away

| Common faults | Likely causes | Remedies |
|---|---|---|
| *Walls* | | |
| Bulging wall | • Overloading<br>• Removal of lateral support<br>• Mortar decay (sulphates)<br>• Through stones/<br>  bricks breaking<br>• Rotting embedded<br>  timbers (see Figure 2.10) | • Engineer or surveyor to<br>  advise on severity and risks<br>• Propping may be needed<br>• Use steel ties<br>• Remove defective material<br>• Repoint or rebuild part<br>  or whole |
| Leaning wall | • Removal of lateral support<br>• Separation from<br>  buttressing wall<br>• Failure of foundation<br>• Thrust from spreading roof | • As above<br>• Reinstate lateral support<br>• Rebuild buttresses<br>• Underpinning as last resort |
| Cracks in walls | • Thermal or moisture<br>  movements<br>• Differential settlement<br>• If above first floor, suspect<br>  deterioration of support, e.g.<br>  bresummer (see Figure 2.11) | • Investigate severity of<br>  cracks, their age and 'live'<br>  or 'dead' by monitoring (see<br>  Part 2.3.3)<br>• Repair or part rebuild<br>• Bricklayer's work |
| Spalling of bricks<br>or stones | • Frost attack (salts)<br>• Expansion when bricks/<br>  stones 'pinched' between<br>  hard cement mortar –<br>  joints cannot 'breathe'<br>• In stone masonry could<br>  be 'face bedded' | • Cut out and replace with<br>  bricks/stones to match in<br>  colour, texture and<br>  permeability<br>• Mortar to be gauged or lime<br>  rich (not cement)<br>• Bricklayer's work |
| Cracks in arches | • Movement in surrounding<br>  wall, probably the<br>  abutment<br>• Deterioration of embedded<br>  timber (see Figure 2.12) | • May not be serious<br>• Engineer or surveyor to advise<br>• Propping may be needed<br>• Needles<br>• Rebuild if severe |

| Common faults | Likely causes | Remedies |
|---|---|---|
| Moss and algae growth | • Any source of dampness particularly leaking rain water systems | • Deal with source of damp, dry out, brush off, treat with a biocide |
| Spalling of render | • Inflexibility of cement-rich render – cracks allow moisture to penetrate – salts or frost or movement cause separation | • Hack off loose render, rake out joints and replace with gauged, lime or hydraulic lime mortar (not cement)<br>• Textured finish (wood float) |
| Loss of pointing in brick or stone masonry | • Old age<br>• Excessive exposure to weather<br>• Sulphate or frost attack | • Repoint if necessary, i.e. when loss allows excessive water penetration. Use mortar never stronger than bricks or stones. See Figure 2.13 |
| Dampness (1) | • Rain penetration through defective gutters, flashings, pipes, roof coverings | • Find source and repair without delay (see roof faults above) |
| Dampness (2) | • Rising damp – absence of damp-proof course (dpc) or failure of early one (pre-1920 felt, perhaps) | • Depends on severity and whether dpc exists<br>• Check ground level<br>• Install ground drainage<br>• Install dpc<br>• Improve ventilation (see Section 2.6) |
| Dampness (3) | • Condensation – moisture in or on surface of cold materials. Can be confused with (1) and (2) above | • Improve ventilation and heating<br>• Insulate 'cold bridges'<br>• Use insulated linings or anti-condensation paints for mild outbreaks |
| Rotting door and window cills or roof facias | • Wet rot due to moisture penetration due to poor specification or lack of maintenance | • Cut out affected parts and replace with treated timber (carpenter/joiner skills needed)<br>• Total replacement as last resort |
| Crumbling internal plaster | • Dampness<br>• Wear and tear<br>• Movement in substrate | • Deal with dampness<br>• Hack off and replace with modern alternative |
| Crumbling cob | • Too damp or too dry<br>• Earth walling normally needs a moisture content 5–10%<br>• Overloading<br>• Impact damage<br>• Rat runs<br>• Impermeable linings | • Avoid use of dpc<br>• Improve ventilation (wall must breathe)<br>• Remove impermeable linings or renders<br>• Get rid of rats and repair<br>• Cob expertise needed |
| Cracking between outside wall and flank wall | • Probably associated with bulge<br>• Overloading | • Engineer or surveyor to advise<br>• May be harmless |

| Common faults | Likely causes | Remedies |
|---|---|---|
| | • Wind suction<br>• Rotting timber<br>• Inadequate bonding<br>• Differential settlement | • Repair using ties or rebonded masonry |
| Cracking and rust stains on concrete | • Carbon dioxide in the air dissolves in water, seeps in and causes 'carbonation'<br>• Steel rusts and expands breaking concrete<br>• Severe in very humid conditions | • Structural engineer to advise on extent of cutting out and replacement<br>• Specialist work only |
| *Floors and ceilings*<br>Suspended timber floor sagging | • Joists and wall plates rotting in a damp wall<br>• Overloading<br>• Weakening by excessive notching for pipework in the past | • Old floors may sag but have settled<br>• Check condition of joists and repair as necessary<br>• Raise level with new joists if needed |
| Excessive vibration of suspended timber floor | • As above<br>• Looseness may be poor construction, e.g. no herring bone strutting<br>• Excessive notching | • Check all timbers for decay and repair as necessary<br>• Insert extra joists<br>• Re-nail loose boards |
| Rotting joists and wall plates in suspended ground floor | • Lack of cross-ventilation<br>• Air bricks blocked<br>• No dpc<br>• Very damp ground/ flooding | • Unblock vents and/or create new ones<br>• Check dpc and repair<br>• Repair timbers with treated timber (see Part 2.8.5) |
| Fragile floorboards | • Commonly 'woodworm' (common furniture beetle)<br>• Encouraged by floor finish that does not 'breathe', e.g. linoleum | • Active infestations need treatment<br>• If severe cut out and replace<br>• Old infestations can be left<br>• Use permeable floor finish |
| Cracked and bulging lath and plaster ceilings | • Floor vibration<br>• Decay of laths or lath nails<br>• Careless past repair work | • Avoid unsightly patching with plasterboard<br>• Remove whole ceiling only if damage severe<br>• Repair ceiling from above (skilled work)<br>• Propping is advised (see Figure 2.14) |
| Dampness or staining of ceiling | • Roof or plumbing leaks from above<br>• Edges of baths and shower trays<br>• Blocked or non-existent overflows | • Deal with source of moisture. Allow to dry out and redecorate using sealing coat if necessary |

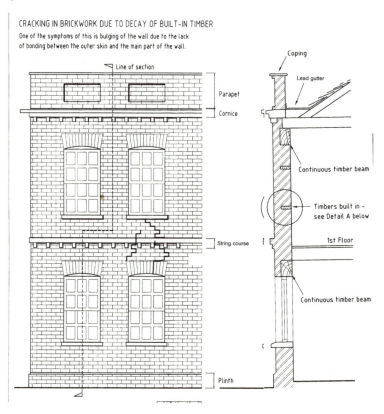

CRACKING IN BRICKWORK DUE TO DECAY OF BUILT-IN TIMBER

One of the symptoms of this is bulging of the wall due to the lack of bonding between the outer skin and the main part of the wall.

Line of section

Parapet

Cornice

String course

Plinth

Coping

Lead gutter

Continuous timber beam

Timbers built in – see Detail A below

1st Floor

Continuous timber beam

## ELEVATION AND SECTION

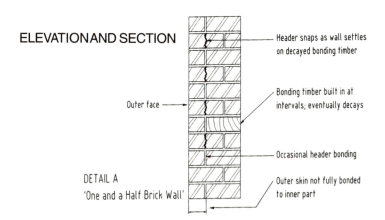

Header snaps as wall settles on decayed bonding timber

Bonding timber built in at intervals; eventually decays

Outer face

Occasional header bonding

Outer skin not fully bonded to inner part

DETAIL A
'One and a Half Brick Wall'

*Figure 2.10   Cracking in brickwork due to decay of built-in timber*

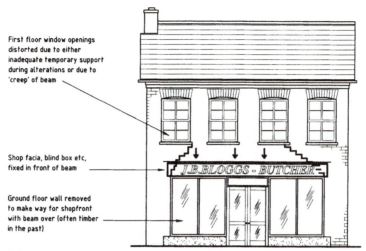

First floor window openings distorted due to either inadequate temporary support during alterations or due to 'creep' of beam

Shop facia, blind box etc, fixed in front of beam

Ground floor wall removed to make way for shopfront with beam over (often timber in the past)

J.B.BLOGGS - BUTCHER

Similar cracking and distortion of openings above is often associated with bay window construction, even when the bay windows are constructed as part of the original building

This type of defect is fairly common where 19th - century houses have been adapted to shops and other 'High Street' uses

*Figure 2.11   Cracking of wall due to movement of members in a wall*

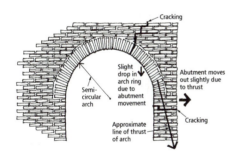

Cracking

Slight drop in arch ring due to abutment movement

Abutment moves out slightly due to thrust

Semi-circular arch

Approximate line of thrust of arch

Cracking

*Figure 2.12   Cracking of arch due to movement of abutment*

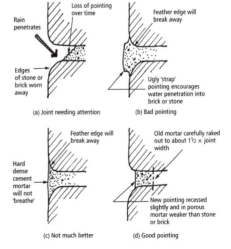

Loss of pointing over time

Rain penetrates

Feather edge will break away

Edges of stone or brick worn away

Ugly 'strap' pointing encourages water penetration into brick or stone

(a) Joint needing attention

(b) Bad pointing

Feather edge will break away

Hard dense cement mortar will not 'breathe'

(c) Not much better

Old mortar carefully raked out to about 1½ × joint width

New pointing recessed slightly and in porous mortar weaker than stone or brick

(d) Good pointing

*Figure 2.13   Good and bad practice in re-pointing old masonry*

| Common faults | Likely causes | Remedies |
|---|---|---|
| *Foundations and ground conditions* | | |
| Settlement of foundation (uniform or absolute) | • Some is inevitable<br>• Long-term compression of clay<br>• Ground water movements<br>• Tree roots<br>• Overloading<br>• Decay of timber foundations<br>• Nearby disturbances<br>• Mining subsidence | • Structural engineer to advise on severity<br>• Dewatering<br>• Remove trees (but beware)<br>• Underpinning as last resort<br>• May have stabilized, then accept it |
| Settlement of foundation (differential) | • As above but non-uniform pattern<br>• Variations in soil condition and bearing capacity | • As above<br>• Much depends on extent of damage to structure above<br>• If stabilized then adapt to it |
| Settlement of solid ground floor slab | • As above but when relatively new building can be failure to compact fill or poor quality fill | • Old tilts may be acceptable<br>• If not and worsening, break out old floor and fill and replace with well-compacted hardcore and damp-proof slab<br>• Alternatively replace with new ventilated suspended floor |
| Unpleasant smell (drains) | • Broken, blocked and leaking drains<br>• Internal manhole cover or gully not sealing<br>• Can be associated with dampness | • Check visible parts of drain system<br>• Have drains tested (specialist or competent builder) and repair or clean as necessary<br>• Internal manholes to have double sealed covers |
| Cellar dampness | • Pre-20th century cellars are usually damp from ground moisture<br>• Could be broken water main or suface water drain | • Check drains and surface water<br>• Reduce ground water level by drainage<br>• Improve ventilation<br>• If absolutely necessary 'tank' cellar with proprietary system (specialist firm advised). See Part 2.6.3 |

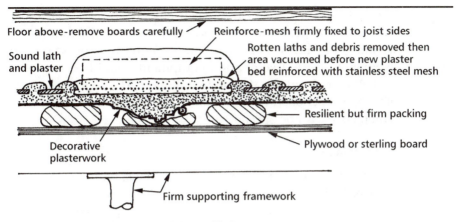

Floor above-remove boards carefully

Reinforce-mesh firmly fixed to joist sides

Sound lath and plaster

Rotten laths and debris removed then area vacuumed before new plaster bed reinforced with stainless steel mesh

Resilient but firm packing

Decorative plasterwork

Plywood or sterling board

Firm supporting framework

Figure 2.14   Repair of a ceiling with decayed laths

## 2.5   Services

### 2.5.1   Electricity

They may be rare but there are still properties in remote areas with no electricity supply. When the property is in a town or village the lack of a supply would almost certainly be no problem; it could be provided at reasonable cost. The cost of supply to an isolated farmhouse in the Yorkshire Dales, however, may well be prohibitive. In the past, petrol powered generators have been used in such places but they require fuel, of course. Solar panels can provide sufficient electricity in the summer months when demand is low. This might suffice if the property is intended for use as holiday lets but there would need to be a supplementary supply in the winter. Solar electric power in Britain has yet to catch on and as a result is relatively expensive to install. Small-scale wind power has much to commend it as far as technology and costs are concerned and it may well achieve a comeback. An installation, however, will normally require planning permission which in view of the stringent planning controls in national parks and some favoured holiday locations would be difficult to obtain.

One has to remember that electricity now provides far more than light and heat. It powers most DIY tools, contractors' plant, washing machines, TV, video, radio, computing and so on. An expansion in the use of such equipment should be expected and allowed for at the time of installation. The sooner the type of supply is chosen and installed, the longer the period over which costs can be spread.

## The electrical installation

Electrical wiring more than 30 years old will often be defective or grossly inadequate for current and future needs. Rewiring by a competent person is the only option. In old properties much evidence can be found of bad and often dangerous wiring carried out in the past. Such conditions must be approached with extreme caution. The prudent surveyor will check that the electricity is turned off at the main before proceeding with a close inspection of the wiring.

Modern cables are generally PVC (polyvinyl chloride) insulated and more efficient than the rubber ones which they started to replace in the 1950s. Old rubber sheathed cables are still to be found in older houses but they will almost certainly have perished and must be replaced. If the old wiring is in conduit it will be very much easier to replace.

The modern way of providing electrical power is via a ring main system running from a 30 amp fuseway on the consumer unit and feeding large numbers of socket outlets. The standard power plug is the familiar three-pin (squared pins) fused type and this has superseded a variety of earlier forms.

The installation and testing of electrical circuits must be done in accordance with the Institution of Electrical Engineers (IEE) regulations. Armed with a 'megohmeter' a surveyor can test a simple installation.

Electrical installations vary enormously even in modern houses. Since the advent of Parker Morris standards (Homes for today and tomorrow: MOHLG) in 1961 the number of socket outlets per room has increased considerably and two- and three-way switching of lighting is now common. More elaborate installations will include spotlights, dimmer switches, heat sensitive security lighting, socket outlets controlled from the door and so on, all of which can result in very much higher costs of installation.

Many owners will derive some assistance from having the electrical installation done by a competent electrical engineer. Others may be attracted by a DIY approach and they will find the guidance provided by the main DIY stores and the various publications available of considerable use, as we shall see in Chapter 10. Safety is of paramount importance when dealing with electricity.

### 2.5.2 Gas

As with electricity, the older the property the more likely it is that a gas installation will be defective, mostly through leaking or blocked pipes and appliances. There can be no question in this case that the supply authority should be called in to test the pipes and appliances. A suspected leak will usually bring a prompt reaction but in the meantime the supply should be turned off at the meter. Most surveyors will be reluctant to test a gas installation apart from commenting on the general condition of the meter, taps and appliances.

Gas is not an essential source of energy but has a record of being cheap and efficient. If there is a supply in the road adjacent to a building, the cost of providing a service to the ground floor is usually waived.

Cooking by gas is popular because the heat is immediately available and easily adjustable. Microwave technology has reduced its importance to some extent.

If there is no possibility of a piped gas supply then bottled gas can be used for both cooking and gas fires but there must be safe and sufficient storage for the cylinders. A central heating gas boiler would require the installation of a large gas container which in most cases would be obtrusive both visually and physically.

### 2.5.3 Cold water

Water supply is of course the most essential of all the services. In urban locations piped mains water is taken for granted and paid for in water rates. In remote rural areas it may be from a private water supply such as a spring in the grounds of another property. The entitlement to such a supply may need to be confirmed by a solicitor.

If there is no supply to a site and no prospect of a mains supply being provided then a supply by tanker could be a solution, or more realistically a well can be dug. Wells, though, are costly and estimates are difficult. Dowsers have been used with some success to detect the presence, depth and even volume of a potential source of supply. However, expense is always relative and costs may be offset by, say, the asking price of the property or the value of the location in other ways.

Where a water service to a property exists, a surveyor's first task will be to locate the stop valve on the incoming main. This is normally outside the boundary but can be hard to find, partly because it should be at least 750 mm from the surface and its cast iron cover may have been removed or paved over. Old service pipes are often made of lead which is no longer acceptable and should be replaced with polythene or copper. At the point where the main enters the building there should be a stop valve and a drain cock so that the system can be emptied while it is being altered or being left unheated for long winter periods. From that point the main should rise, preferably not adjacent to a cold external wall, to a storage cistern in the roof with branches off to a fitting which requires drinking water such as a kitchen sink. Stop valves should be fitted at the cistern and below the sink to enable tap washers to be changed and other maintenance work. No drinking water should be taken off a storage cistern. See Figure 2.15.

All water pipes should be checked for leaks where visible, or dampness where concealed, together with all taps, valves and so on. Where pipes are lagged, in a roof for instance, the lagging should not be disturbed unless it can be confirmed that it is not asbestos. If in doubt a specialist firm should be called in to inspect the lagging and deal with its removal. This will take time and the roof space will be 'out of bounds' pending the removal of the asbestos.

## 2.5.4  *Hot water and heating*

Old buildings may still have 'direct' systems of heating water but many have been replaced with 'indirect' systems. In the

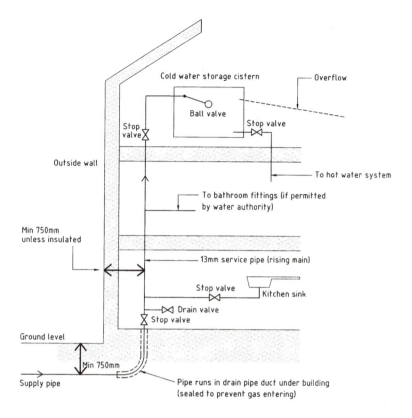

Figure 2.15    A simple cold water system

direct system the water is heated in a boiler and rises by convection to a hot water cylinder usually in a cupboard on an upper floor. As hot water is drawn off to fittings such as baths and basins it is replaced with cold water from the storage cistern in the roof. See Figure 2.16.

Indirect systems are capable of providing central heating with radiators and are less likely to cause furring problems in hard water areas. This is because the hot water that is used is heated in an indirect hot water cylinder by a separate circuit heated by the boiler. This so-called primary circuit requires a small feed and expansion cistern usually alongside the cold water storage cistern in the roof space. See Figure 2.17.

In recent years it has been possible to dispense with cold water storage cisterns by the use of 'combination' boilers. These provide instant hot water without storing it and can feed a

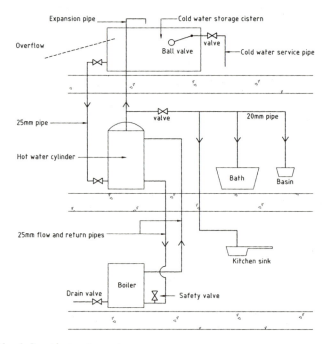

*Figure 2.16   A direct hot water system*

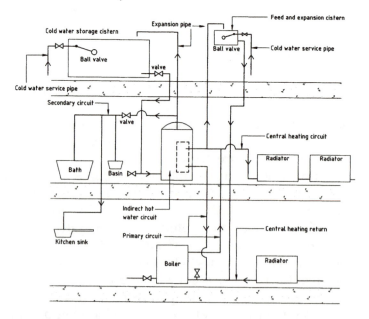

*Figure 2.17   An indirect hot water and heating system*

limited number of radiators. They work on a principle not unlike that of the old instantaneous gas heaters (now rare) but are very much more efficient. Combination boilers have been criticized for producing a slow feed of hot water, particularly in cold weather, so prospective buyers should check that they are adequate for their use, particularly if showers are to be installed. See Figure 2.18. Combination boilers which are also 'condensing' boilers are even more efficient although initially more expensive. They will become readily available in the UK in 1998.

Modern boilers are either solid fuel, oil or gas-fired, the latter being the most popular. Solid fuel and oil-fired boilers require a suitable flue and storage space for the fuel. In the case of oil this usually means the installation of an external oil tank. Gas-fired boilers, which can be floor standing or wall hung, can be connected to a conventional flue but they can also have a balanced flue which is positioned in the wall behind the boiler. This gives a great deal of flexibility in positioning the boiler and it is now common practice for it to be installed in the kitchen. The Building Regulations (Part J) include complex and stringent rules regarding boiler installations, particularly with regard to the following:

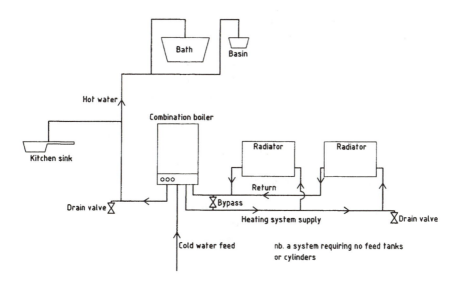

Figure 2.18   A system using a combination boiler

- The provision of an adequate supply of air for combustion and the proper functioning of the flue. This means that either the boiler must be a 'room-sealed' appliance or a ventilation opening must be provided to outside air.
- The installation of a suitable flue. This means that a balanced flue or a suitable fluepipe or chimney must be provided. Old chimneys may require lining and there are proprietary methods of doing this. Flues should have openings for inspection and cleaning.
- Solid fuel or any appliance that gets hot and is likely to heat the floor below it above 100 °C will need a non-combustible hearth; nearby walls may also be affected.

The correct functioning of an existing heating appliance or the selection of a new one are matters which should be dealt with by a heating engineer or a specialist sub-contractor. Where gas is concerned, British Gas will advise or a member of the Council for Registered Gas Installers (CORGI) should be called in. In fact it is a legal requirement that all new gas installations must be carried out by a firm registered with CORGI. Such firms, generally listed in *Yellow Pages*, usually deal with oil and solid fuel installations as well.

## 2.5.5 Drainage below ground

There are two kinds of drainage systems: foul water and surface water. A foul drain carries the waste from WCs, baths, basins, sinks and so on. A surface water drain carries rain water from roofs and paved areas. In some cases the two systems may be combined, with both discharging into a main public sewer, but many authorities will not accept this. Normally the alternative means of disposal are as follows.

### Foul drain: connected to main public sewer

This is generally located in the highway adjacent to the property although in certain cases such as a Victorian terrace of houses the drainage from each house may be collected in a joint pipe before running into the main sewer. See Figures 2.19(a) and (b). New connections to public sewers must be approved by the local authority who will insist on the correct method being used.

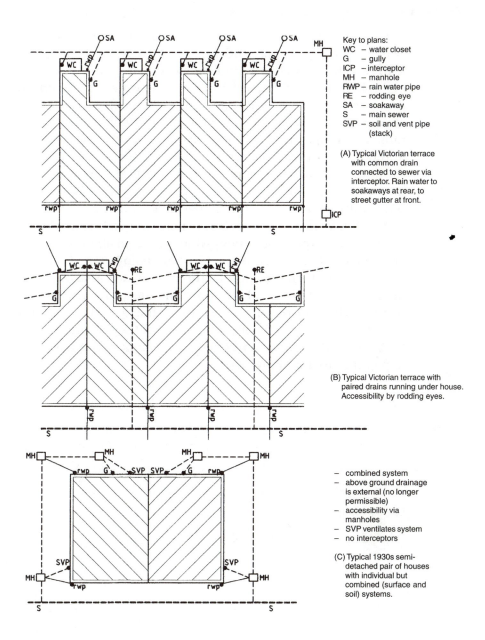

Key to plans:
WC — water closet
G — gully
ICP — interceptor
MH — manhole
RWP — rain water pipe
RE — rodding eye
SA — soakaway
S — main sewer
SVP — soil and vent pipe
(stack)

(A) Typical Victorian terrace
with common drain
connected to sewer via
interceptor. Rain water to
soakaways at rear, to
street gutter at front.

(B) Typical Victorian terrace with
paired drains running under house.
Accessibility by rodding eyes.

– combined system
– above ground drainage
is external (no longer
permissible)
– accessibility via
manholes
– SVP ventilates system
– no interceptors

(C) Typical 1930s semi-
detached pair of houses
with individual but
combined (surface and
soil) systems.

Figure 2.19    Typical drainage layouts

### Foul drain: discharge to a septic tank

This is normally used where main drainage is not available. Waste is conveyed to an underground tank which is designed with two chambers to separate out solid matter and allow it to be broken down by bacterial action. The water or effluent is drawn off to a land drain, soakaway (see below) or nearby stream. Every one or two years the sludge from the tank itself must be pumped out, a service usually provided by the local authority engineering department. Access must be provided for the tanker to reach the septic tank, with its flexible pipe usually not more than 30 m distant. Old septic tanks were usually brick built and are rarely still efficient; particularly they are no longer watertight. Modern septic tanks are usually prefabricated bottle or cylinder shaped containers which are available in various sizes, efficient and easy to install. Their installation should be in accordance with the manufacturer's instructions and to the approval of both the local authority and the local environment agency who will require assurance that adjacent ground conditions are suitable for the disposal of the effluent. Figure 2.20 shows the Klargester type of septic tank which has been in popular use for many years.

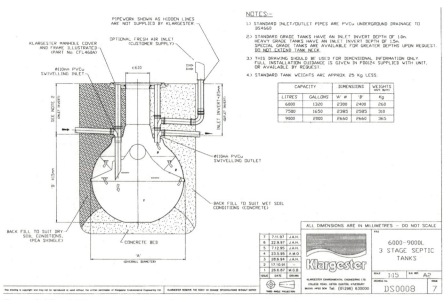

Figure 2.20   A Klargester septic tank

### Foul drain: discharge to a cesspool

Where there is no main drain and no means of disposing of the effluent from a septic tank, a cesspool may be a last resort. This is a large underground tank which collects the sewage and does not treat it. It should therefore be emptied more frequently than the septic tank. Old cesspools may be little more than holes in the ground and may have been leaking for years. Repairs are very difficult and replacement with a modern prefabricated tank, which can be carried out in less than a day, is the preferred option. Where ground conditions require a more robust form of construction, cesspools can be built of precast concrete rings but the sheer volume required (at least 18 000 litres or 18 m³ for one dwelling) means that several interconnected chambers may be necessary. As with septic tanks, the authorities must give their approval to cesspool installations.

### Surface water drain: connected to main public sewer

This is only permissible where the drainage authority operates a combined system. The main system must be capable of carrying the discharge from heavy storms as well as foul sewage. Figure 2.19(c) shows a combined system for a pair of semi-detached houses.

### Surface water drain: discharge to soakaways

Except on sites where the ground is poorly drained or waterlogged, this is now common practice. Old soakaways are often simple shallow holes in the ground filled with rubble which may well be silted up, leading to very damp ground and even flooding in the vicinity. Oddly it is not until they cease to function properly that their location becomes apparent. New soakaways may be of similar construction to old ones except that care should be taken to ensure that they are deep enough and that they are filled with good quality granular material such as broken stone. Alternatively a more efficient soakaway can be built using perforated precast concrete rings or brickwork with some open joints so that water can percolate through to the surrounding soil. This kind of soakaway can be fitted with an inspection cover so that it can be desludged from time to time.

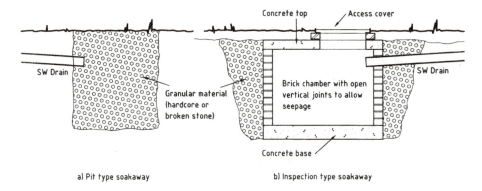

Figure 2.21    Two types of surface water soakaways

This extends its life but it is of course more expensive to build in the first place. See Figure 2.21.

Having noted the positions at which soil and waste pipes enter the ground adjacent to a building, the surveyor's next task will be to locate the covers of inspection chambers, lift them and note the nature, condition and direction of the channels in them. In many pre-twentieth-century properties, however, there may be no such inspection chambers and those that exist may be buried below paving or soil. In such a case it may be difficult to convince the local building control officer that the drain can be used for, say, a new extension unless it is tested (see below). If it fails the test because it is broken or just worn out then the preferred option is to lay a new drain rather than repairs which are usually impracticable and expensive.

### Points to look out for in the drain survey

- Tree roots are attracted to broken drains and can cause irreparable further damage – beware trees growing over drains.
- Old inspection chambers (manholes) should be as watertight as the drain itself – preferably in hard bricks with flush cement mortar pointing, not rendered. If rendered look out for pieces broken by frost action and falling into the channels, causing blockage. The 'benching' may also be defective. See Figure 2.22.
- Foul drains must be ventilated. Check that a ventilated stack pipe has been fitted rising from the 'top end' of the drain to a point above the roof of the building. Most installations in

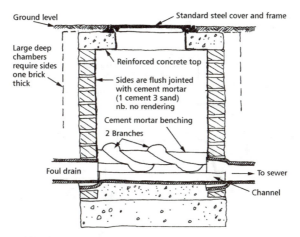

Figure 2.22 *A typical brick inspection chamber (manhole)*

the twentieth century have the familiar soil and vent pipe (SVP). In alteration work where a ventilated stack is difficult to provide, the use of an 'air admittance valve' is now permitted but only inside the dwelling, not outside or in the roof space.

● Old inspection chamber covers and their frames may have deteriorated badly and more recent ones may become distorted in use. To carry the load of a vehicle the cover must be 'heavy duty' and set on a substantial base. Most builders' merchants stock a selection of covers for different situations.

● 'Single stack' above ground drainage where all soil fittings discharge into one vertical ventilated stack is a common feature of modern housing. Alternatively, and certainly in older properties, the waste from baths, basins and kitchen sinks may discharge into gullies which have traps to prevent the escape of foul air. These need careful inspection and where broken or cracked they should be replaced with back inlet gullies which are now the acceptable alternatives. See Figure 2.23. Where such gullies are used internally they must be of the sealed type.

### Drain tests

As discussed earlier, the simplest test of a drain is to release water from fittings and observe its flow through the inspection chambers. This will give a good indication of whether the drain works or not but will not necessarily reveal faults in

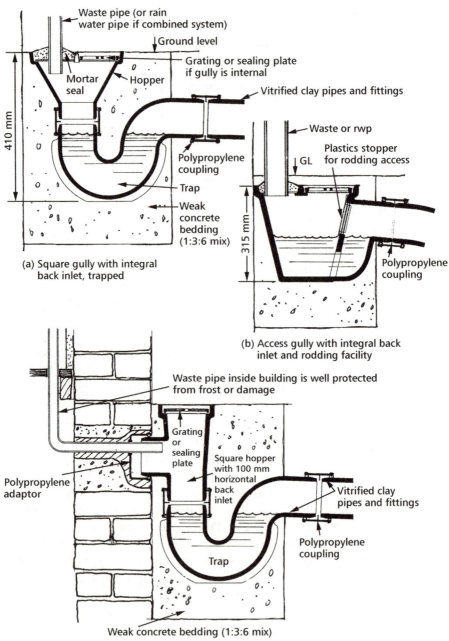

(a) Square gully with integral back inlet, trapped

(b) Access gully with integral back inlet and rodding facility

(c) Gully with horizontal back inlet for waste pipe - the better solution

*Figure 2.23   Alternative gullies (Hepworth Supersleve type)*

the pipes. Further tests may be necessary and some are briefly described here:

- *Rodding* – many contractors and certainly specialists have a set of flexible rods which are used not so much to test drains but to locate blockages and clear them. A set of rods costs between £50 and £200. Inspection chambers and 'rodding eyes' are the means of access for rodding and are a feature of modern systems. Hidden soakaways can be located by rodding surface water drains until an obstruction is met.

- *Water test* – a simple method whereby the drain is plugged at its lower end and then filled with water right up to the top of the inspection chambers if necessary. Suitable plugs are available from most builders' merchants. The level of the water is checked after about two hours and a pronounced loss of water will indicate that the system leaks. Allowance must be made for absorption by the drain materials. One difficulty with the method is that quite high water pressures can be generated and an old drain which functions well under normally low pressure will fail the test. The tester should not over-fill the system in these circumstances and discretion should be exercised in the interpretation of results.

- *Closed Circuit Television (CCTV) survey* – a service offered by specialist firms such as Dyno-Rod or Seerseal. Tiny TV cameras are pulled through the drains and what they see is shown on a nearby monitor. Blockages, breakages, tree roots and so on can be located precisely but skill is required in operating the equipment and it can cost several hundred pounds to test one domestic system. Nevertheless it is a very effective way of testing drains without damaging them. Firms offering CCTV are well represented in *Yellow Pages* and local authority engineering departments may also have the equipment.

## 2.6   Rising damp

### 2.6.1   *The nature of rising damp*

Rising damp occurs when moisture in the ground is drawn into the walls and ground floor of a building by capillary action. Old buildings built with porous materials such as stone or brick in lime mortar are very likely to be affected, particularly on damp sites. The water which is sucked through the pores in the wall or floor may be ground water but a defective rain water system or raised ground levels can allow water to accumulate close to the building instead of draining away.

The height to which moisture will rise in this way depends on the porosity and permeability of the wall and the moisture pressure from the ground. It is usually between about 600 mm and 1500 mm from the base. Loose wallpaper is often a sign. Good ventilation on both sides of the wall will help to keep the level down because it hastens the evaporation process. The top level of damp usually fluctuates with environmental changes and when it recedes it is not unusual to find salts crystallizing on the surface it leaves behind. These soluble salts occur to varying degrees in most building materials and some soils and may be harmless. Where they become concentrated within the fabric of a wall, however, they cause damage by crystallization in the pores and breaking the material up. The damage may be so severe that extracting the salts from a structure by a specialist may be more important than dealing with the damp, although ideally of course both should be treated at the same time. Apart from the physical damage that salts can cause they are also hygroscopic, which means that the wall or floor affected will take up moisture from the air and so never dry out.

In some cases, particularly in unheated old buildings, condensation can be mistaken for rising damp. The fact that only the lower half of a wall seems damp may be the result of moist air condensing on a low cold wall not well ventilated, whereas warm damp air rises to become dispersed at the upper level where the walls are somewhat warmer.

## 2.6.2 *How to deal with rising damp*

A surveyor will give advice on methods of dealing with rising damp, having first measured the moisture content and salt content of the building using specialist equipment such as a 'protimeter' kit. Advice can also be had from one of the many specialist firms dealing with damp and timber infestation. These are listed in *Yellow Pages*.

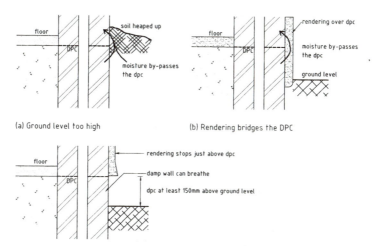

(a) Ground level too high          (b) Rendering bridges the DPC

(c) The ideal arrangements – but note that external
masonry below dpc level must be frost resistant

*Figure 2.24    Bridging of a dpc in a cavity wall*

An owner should regard such advice with a certain amount of circumspection. People are often persuaded, under pressure from mortgagors such as building societies and specialist firms, that a damp proof course (dpc) should be installed when it is not really necessary and not likely to be effective. Certain measures should be tried first, such as:

- Improve heating and ventilation, particularly if there is a history of poor provision in either case.
- If there is a damp proof course check to see whether it has been bridged. Figure 2.24 shows how this can happen with soil or rendering.
- Check that the wall is not sealed – it must breathe.
- If the soil adjacent to the building seems excessively damp, investigate the possibility of draining it. The lie of the land will dictate whether this is possible. It is a problem with which many contractors are familiar and have the equipment to carry out. Figure 2.25 is an illustration of the technique. The objective here is to reduce the ground water level, ideally to a level below the foundations but not so far below that soil shrinkage occurs. In the case of a historic building no digging should take place without the knowledge of the local conservation officer and archaeological group.

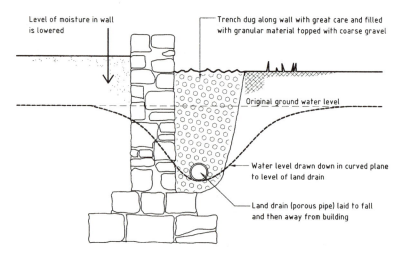

Level of moisture in wall is lowered

Trench dug along wall with great care and filled with granular material topped with coarse gravel

Original ground water level

Water level drawn down in curved plane to level of land drain

Land drain (porous pipe) laid to fall and then away from building

*Figure 2.25    Drying trench method of reducing the level of rising damp*

If these measures appear to have no effect after a reasonable time, the insertion of a dpc should be considered. Again, though, a note of caution should be sounded. Very thick old walls, piers and chimney breasts are notoriously difficult to damp proof using any of the standard methods available.

Some of these methods are listed and briefly described here. Their suitability in any particular case should be discussed with a surveyor or specialist installer.

### Insertion of strip dpc

This is extremely specialized. It involves the use of a special chainsaw to cut out a horizontal joint in a masonry wall and insert in its place a dpc, traditionally of slates or copper sheet but more recently a polyester or epoxy resin liquid which fills the joint and then sets hard. This method will only work satisfactorily where there are continuous horizontal mortar joints as in brickwork or ashlar masonry. Neither would it be suitable for party walls unless the adjoining owner is in full agreement. Nevertheless it has been successful, although costly. See Figure 2.26.

### Electro-osmosis

This works on the principle that moisture in porous material will travel from a positively charged element (anode) to a

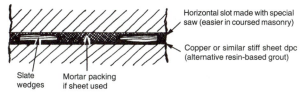

Figure 2.26    Insertion of a strip dpc

negatively charged one (cathode). There are two systems, a passive one and an active one. In the latter case a permanent electrical supply is needed. The electro-osmosis system is relatively easy to install and some success is claimed for it where the dampness is not too severe. However, there have been problems with it, particularly when the elements have been broken or disturbed by alterations, and for this reason specialist firms will generally recommend an alternative.

## Capillary tubes

Also known as 'knapen' tubes or 'atmospheric siphons' and manufactured for many years by Doulton in Staffordshire, these small ceramic tubes are porous. When positioned in short boreholes at regular intervals along a wall they increase the rate of evaporation from the wall and so reduce the moisture level rather than eliminate it. It has been found, though, that after years of use salts are drawn into the tubes where they crystallize and block the evaporation, so even though they can be replaced they are no longer considered very effective. See Figure 2.27.

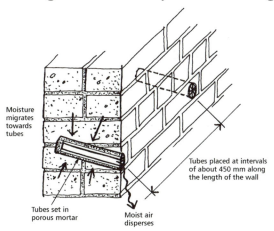

Figure 2.27    Porous ceramic tubes as an aid to evaporation

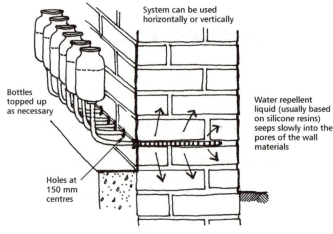

*Figure 2.28   Chemical infusion as a dpc*

## Chemical systems

This is the most common method of dealing with rising damp. Silicones or silanes in solution are injected at intervals of about 150 mm into the wall where they harden and seal the pores in the masonry. The liquid is said to be injected when it is pumped and infused when it is placed by gravity. The success rate of this method is high, although much depends on the care and skill of the operator. The extent to which the injected solution has filled all the pores in an old rubble wall, for example, can be very uncertain. See Figure 2.28.

## Mortar injection

This process involves drilling 22 mm holes about four-fifths of the way through a wall at 225 mm centres. Using a caulking gun, the holes are filled with a mortar composed of cement, quartz, organic acids and water which reacts with the moisture in the wall to provide not only a barrier to rising damp but also additional strength to the wall. One advantage claimed for this method is that the mortar remains active when fully cured. When moisture reappears the chemicals in the mortar will reactivate.

Specialist firms which use these methods of damp proofing will usually require the walls to be stripped of plaster up to the original damp level and to be replaced with a salt-retardant alternative. This is to ensure that the salts in the wall, having solidified as the wall dries out, cannot continue to attract moisture from the air and cause further deterioration of the wall. The walls should be tested for salt content to determine whether such replacement plastering is necessary because the process can be very disruptive and the very hard, non-porous plaster used can be of doubtful benefit because it does not allow the wall to breathe. In some cases, just leaving the wall unplastered for enough time so that the salts can migrate to the surface and be brushed off will suffice, but unfortunately this is rarely possible.

### Wall linings

Where none of the methods so far described seem appropriate, hiding the problem behind some kind of wall panelling should be considered, as it has been for centuries in the past. Of course, only the unsightliness of damp walls is being dealt with in this way and there are serious problems with it, such as preventing the material of the lining from deteriorating while at the same time allowing evaporation to occur. The simplest form of wall lining is plasterboard on softwood battens carried either to the top of the wall or part way up as in the traditional dado. Plasterboard, though, will deteriorate rapidly in damp conditions and so will the battens even when treated, so a polythene membrane should be used against the wall. This will prevent any evaporation from the wall and the resultant build-up of moisture may lead to damp patches at a higher level. An alternative method would be to fix to the wall a proprietary ventilated sheet material such as 'Newlath' or 'CDM Zipseal' which allows the wall to breathe but still provides a key for a new plaster finish. Ventilation of the cavity behind the lining is achieved by leaving gaps at the top and bottom of the wall. See Figure 2.29. Newlath is a product of John Newton & Co. Ltd, 12 Verney Road, London SE16 3DH, tel: 0171 237 1217. CDM Zipseal is a product of Safeguard Chemicals Ltd, Red Kiln Close, Redkiln Way, Horsham, Surrey RH13 5QL, tel: 01403 210204/210648.

## 2.6.3 Damp basements

Damp basements present special problems. The amount of moisture present will govern the damp proofing method to be

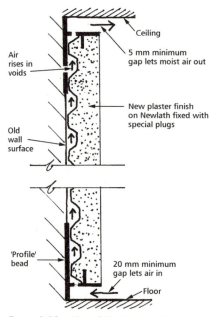

Air rises in voids

Ceiling

5 mm minimum gap lets moist air out

New plaster finish on Newlath fixed with special plugs

Old wall surface

'Profile' bead

20 mm minimum gap lets air in

Floor

Figure 2.29 Use of Newlath ventilated lining of damp walls

used. Moderate dampness may well be alleviated by improving heat and ventilation but where the basement floor is below the water table it may be necessary to line the walls and floor with a jointless membrane – a process known as 'tanking'. Simple screeds and plasters are rarely able to cope with the level of pressure that ground water can exert, so it is usually necessary to use a proprietary system. Materials such as Newlath and CDM Zipseal are well suited for use on both the walls and the floor of a damp cellar and offer the advantage that they can be dry lined afterwards. Alternatively, wet systems such as that marketed by Sika Ltd have been used successfully for many years. The Sika system comprises several thin coats of a dense cement mortar containing a waterproofing compound. The number of coats depends on the water pressure and the condition of the structure. The whole operation can take two or three days. Where free water is likely to build up behind a damp-proof membrane it may be necessary to dispose of the water by channelling it to a sump from which it can be pumped to ground level. A submersible pump has a float switch which only operates when sufficient water has been collected in the sump. See Figure 2.30. Sika Ltd are at Watchmead, Welwyn Garden City, Herts AL7 1BQ, tel: 01707 394444.

## 2.7    Foundations and ground conditions

### 2.7.1   The approach

As suggested earlier, the cracking and movement of a building's structure are often caused by the failure of its foundations. If

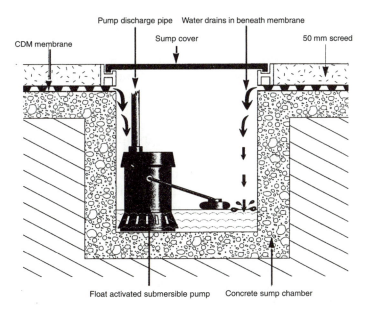

Figure 2.30  Sump with submersible pump

such a failure has already occurred it will be necessary to investigate the cause in order to suggest remedial action. If the structure appears to be sound, however, but a conversion or alteration is planned then the foundations should still be investigated to ensure that they have the capacity to carry any additional or adjusted loads which may result.

Since it is possible to expose only a small part of an existing foundation to inspect it, as much information as possible needs to be obtained by other means. For example:

- Old drawings – are any available which show the work below ground? The older the building the less likely that this will be the case.
- Local literature – historic documents may include specifications and builders' accounts for past work.
- Geological maps – may be helpful on a broad scale in identifying underlying conditions and local characteristics but professional interpretation is often needed.
- Local experience – practitioners, local authority officers, contractors and others who have experience of ground work in the locality may give advice although it must be borne in

mind that considerable variation in ground conditions can occur even in a small area.

- Looking at the building as a whole and its neighbours to detect structural cracking and distortions.

### 2.7.2  What to look out for

It pays to be wary of certain ground conditions which may exist on new or old sites. They rarely prevent development but they may well affect the design and construction of a new foundation or have caused an old foundation to fail. Some of these conditions are listed here with a brief note of explanation. Any one may be serious enough to justify an investigation by a specialist engineer.

- *Clay soils* – 'shrinkable' clays such as London or Oxford clay will swell and shrink as their moisture content changes, thereby causing considerable movement in any foundation on which they rest.
- *Trees* – fast-growing deciduous trees such as poplar, sycamore and willow will cause seasonal variations in the ground water level which will exacerbate the clay problem outlined above. In new developments trees should not be too close to buildings unless special foundations are used. See Figure 2.31.

- *Mining subsidence* – generally less of a problem than it used to be and only in certain areas but old mines may still cause settlement. This is a case for a specialist scientific survey.
- *Land slips* – typically a slow downward movement of clay soil on a sloping hillside made evident by leaning trees and distorted buildings. Damage to a building is almost inevitable.
- *Swallow holes* – occurring in chalk and limestone rocks,

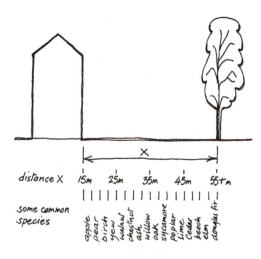

Figure 2.31   Trees and buildings

these are cavities formed by moving water washing away the fine material in the rocks. Calcium carbonate ($CaCO_3$) is soluble in water. Sudden collapse of the rock above the cavity may well induce a sudden failure of a foundation above it. Another case for the specialist survey.

- *Underground water* – wet soil is generally weaker than dry soil and may exist in an unstable condition so that when disturbed or subjected to pressure changes it will move. Sand is particularly susceptible in this respect. Movement in the soil will precipitate a movement in a foundation.
- *Pollution* – a phenomenon of the post-industrial age, the contaminated site is all too common. Local historic records should provide a clue as to what contaminants are present and specialist advice is available as to the hazards involved and ways of dealing with them. The local environmental health officer should be the first person to contact.
- *Natural gases* – radon is a radioactive gas which occurs in the West Country, Northamptonshire and Derbyshire. New buildings in these areas will require radon-proof barriers to be incorporated in the design but existing buildings are difficult to treat in the same way. The advice of the local building control officer should be sought. Methane is another naturally occurring gas although it is more often associated with areas of landfill where it is a product of decomposing organic material. Methane is flammable and toxic and expert advice is needed to deal with it.
- *Made ground or fill* – may be very old in which case it will be of archaeological interest or it may be recent in which case it may contain not only natural soil but waste materials as well. In either case a good deal of organic material may be present and it can be poorly compacted, rendering it unsuitable for simple foundations.
- *Soils susceptible to freezing* – wet silty soils close to the surface and in frost prone areas may freeze and swell, shrinking again when it thaws. In the British Isles this 'frost heave' has little effect below a depth of about 500 mm but it is one reason why new foundations are normally placed at a depth of at least 1 metre. Another reason is to avoid clay shrinkage as mentioned above.

### 2.7.3  Trial holes

One of the prerequisites of any new building scheme is the soils investigation which will vary in method and scope depending on the size of the proposed building and the nature of the site. Soils investigations reduce the risk of unforeseen conditions appearing once construction has started and it is a matter of judgement for the owner and design team whether the cost of a soils investigation can be justified in order to reduce that risk. It is not possible to remove the risk altogether, only to reduce it. Generally speaking, a comprehensive borehole investigation and report carried out by specialist engineers will be required for large or medium-sized projects. Digging trial pits with a small excavator, however, usually provides all the information needed for a small new build or alterations contract provided it is done to a reasonable depth. For simple house foundations, for example, a trial pit should be a least 2.5 metres deep (within the scope of, say, a JCB Sitemaster) and wide enough for someone to climb down on a short ladder. See Figure 2.32.

Where it is necessary to investigate the foundations of an existing building, as in the case of conversions, adaptations or extensions, great care must be exercised in digging trial holes adjacent to the existing foundations which must not be disturbed. Holes about 1 m wide and not closer than 5 m apart may be permissible but much depends on the condition of the building and the depth of the foundation uncovered. This kind of work is best left to the experienced contractor who will have on hand any temporary supports which may prove necessary.

Whatever the form of trial or boreholes adopted, they should be quickly examined by those who have an interest (building control officer, surveyor, contractor, engineer and so on) and back-filled

**Note**
- Do not dig until inspectors are ready
- Beware risk of collapse - use strutting in poor soils
- Backfill immediately after inspection
- Depth depends on foundation type, building load and anticipated soil conditions but not less than 2 m
- Width sufficient to allow inspection of the sides

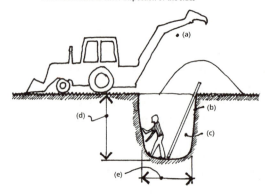

Figure 2.32    Digging a trial hole

immediately for reasons of safety. The information to be gleaned from the inspection will usually include the following:

- *Depth of top soil if any* – a valuable resource which should not be wasted or built over; stacking it for later reuse requires space and will incur a cost.
- *Depth of strata below top soil* – this will determine the depth at which foundations can be placed.
- *Type of subsoils* – to estimate their bearing capacity, the likelihood of movement (as in clays already discussed), the ease with which it can be dug out (a matter for the estimator) and the extent to which temporary strutting may be required (a matter for the general foreman).
- *Presence of sulphates* – a sample of soil can be taken to the laboratory and tested for the presence of such salts as magnesium, sodium or calcium sulphate which when dissolved in ground water will attack the ordinary portland cement in concrete.
- *Presence and depth of water table* – when high it may affect the strength and likely movement of soils, render sulphates more dangerous, increase risk of leaking basements and require pumping out when construction is under way, another item for the contractor to price.
- *Presence of made ground* – as referred to above – this is not always detectable to the untrained eye, particularly if it is very old, but the 'feel' of the soil, its texture, colour and smell, together with the odd piece of broken crockery or brick, will sometimes lead one to suspect that it has not been naturally deposited.

## 2.7.4  Some foundation problems

All buildings settle into the ground during and immediately after construction. This short-term settlement is a result of the soil being compressed and is inevitable except when building on rock. In the case of clay soils, though, long-term consolidation, usually the squeezing out of water, will give rise to settlement occurring for a considerable period after construction. As we have seen, even after that movement can occur when the moisture content of the clay changes. Gravelly soils settle more quickly and are more likely to stabilize permanently.

If the settlement of a building is uniform it may cause little damage except to rigid connections to it such as drain pipes. Uneven or differential settlement, on the other hand, almost invariably causes distortions and cracking in the structure of the building. Figure 2.33 shows some typical effects of differential settlement in a two-storey dwelling.

We have seen how the type and condition of the soil below a foundation may cause settlement. Other causes may stem from the use or construction of the building itself. These include:

- *Inadequacy* – foundations of any age may have been built too narrow or too shallow for their required function. Very good bearing soil may have compensated for this for a very long time but as soon as conditions change in the ways outlined below, the weakness of the poorly built foundation may become manifest.
- *Overloading* – this may occur in three ways. Firstly, the user of a building may be subjecting it to far greater loads than it was originally intended for. A suspended domestic floor, for example, is rarely built to carry heavy machinery, large

If different parts of a foundation settle by different amounts, distortion and cracking of walls will occur. At (d) the ends have moved more than the centre ; at (e) the opposite has occurred. Where there are openings (windows and doors) in a wall, cracking will often start from the opening. At (f) a low extension has settled differentially from the main building.

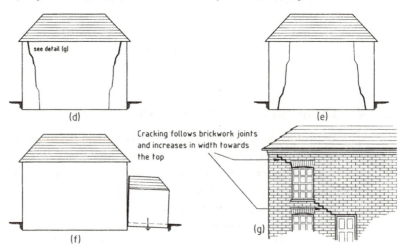

*Figure 2.33  Differential settlement*

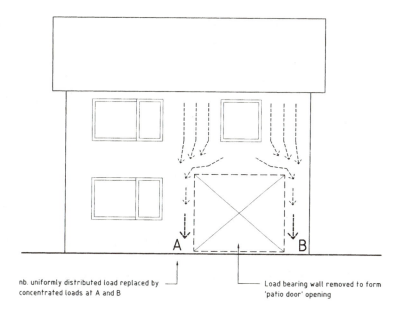

nb. uniformly distributed load replaced by concentrated loads at A and B

Load bearing wall removed to form 'patio door' opening

*Figure 2.34    Redistribution of loads caused by forming a new opening*

water tanks or Aga cookers. Secondly, alterations may have taken place in such a way that loads that were once uniformly spread are now concentrated. Forming new openings in old walls may have this effect as shown in Figure 2.34. Thirdly, the addition of another storey such as a first floor extension or a loft conversion will increase not only the dead load but also the live load (more furniture, people, books, etc.) on the original foundations. Old cottages are particularly susceptible to this effect. In all these cases, the overloading may have little effect on a well-built foundation on a good bearing soil but an undetected weakness in either will probably lead to differential settlement.

● *Loss of bracing* – the foundations of an old building are often braced by the stiffening effect of the walls built on them. By removing such walls or making large openings in them there is a danger that the foundations will be relieved of pressure in a non-uniform way and subjected to ground heave with the resultant breaking.

● *Deterioration of material* – by the action of water, contaminants, acids and salts such as the sulphates already discussed, foundations can be attacked and eventually

79

fragmented. Steel, concrete, timber and limestones may all be affected, particularly at shallow depths where corrosion is caused by alternate wetting and drying.

- *Undermining by adjacent works* – where construction work takes place next to the building there is a risk that a foundation will be undermined. Unless the disruptive work is far enough away from the foundation to avoid interference with its 'zones of influence', some form of underpinning may be necessary. See Figure 2.35 and the next part on underpinning.

### 2.7.5  *A note on underpinning*

There are a number of techniques for underpinning buildings and they all have one objective, i.e. to transfer a foundation load from its existing level to a lower one. This is usually because:

- settlement has occurred or is occurring, or
- greater load capacity is needed, or
- the adjacent ground is about to be lowered, as in Figure 2.35.

The traditional approach is to use 'continuous underpinning'. In this method the work is carried out in sections so that only

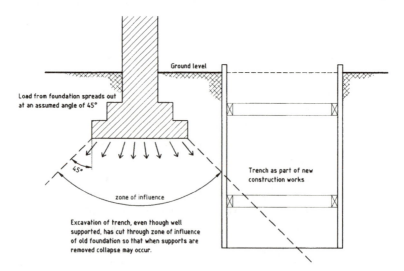

*Figure 2.35   Digging too close to an old foundation may cause it to collapse*

a small part of the building is unsupported at any one time. Holes are dug to a level well below the old foundation and a section of new foundation, usually in concrete, is built below it. The difficulty of this method is that it is very difficult to avoid some settlement of the building while the work proceeds. It is also a lengthy labour-intensive process. Alternative methods make use of reinforced concrete beams, hydraulic jack and piles, and while the success rate is high they can be very expensive and disruptive. See Figure 2.36.

There is a tendency for some people, often professionals, to recommend underpinning as a cure for all foundation ills, discounting any question of cost or the risk that more harm can be done than good. Underpinning should be regarded as a last resort, to be used only when the cause cannot be eradicated, where the settlement is continuing and is causing irreparable damage to the building's structure. The mere fact that a building is moving slightly is insufficient justification for such an extreme measure as underpinning to be adopted.

When a building owner suspects that underpinning may be necessary he or she should:

- consult a structural engineer who has a good record of dealing with existing or old buildings;
- question the extent to which the settlement is 'alive' or historic;
- ascertain the cause of settlement and appraise remedial measures;
- evaluate the damage caused by the settlement;
- evaluate the estimated cost of underpinning, the disruption and the risks involved.

## 2.8 Timber decay

### 2.8.1 The importance of moisture

Certain species of fungi and beetles attack wood when it is damp. Both hardwoods and softwoods can be affected although the

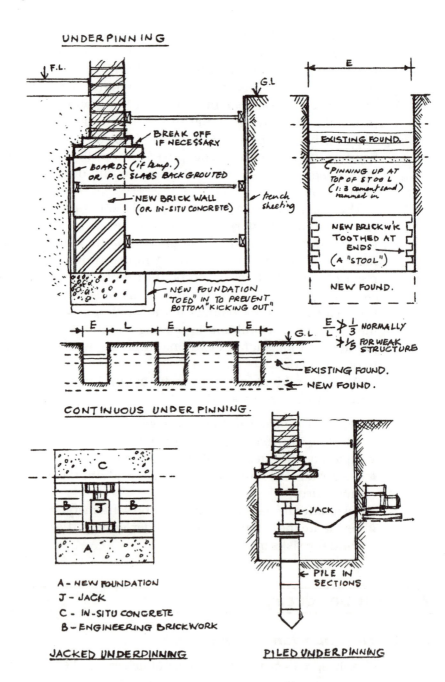

UNDERPINNING

F.L.

G.L

BREAK OFF
IF NECESSARY

BOARDS (if temp.)
OR P.C. SLABS BACK GROUTED

NEW BRICK WALL
(OR IN-SITU CONCRETE)

trench
sheeting

NEW FOUNDATION
"TOED" IN TO PREVENT
BOTTOM "KICKING OUT".

E

EXISTING FOUND.

PINNING UP AT
TOP OF STOOL
(1:3 cement sand)
rammed in

NEW BRICKW'K
TOOTHED AT
ENDS
(A "STOOL")

NEW FOUND.

E  L  E  L  E  L  E

G.L

$\frac{E}{L} \not> \frac{1}{3}$ NORMALLY
$\not> \frac{1}{5}$ FOR WEAK
STRUCTURE

EXISTING FOUND.

NEW FOUND.

CONTINUOUS UNDERPINNING.

C

B   J   B

A

A - NEW FOUNDATION
J - JACK
C - IN-SITU CONCRETE
B - ENGINEERING BRICKWORK

JACKED UNDERPINNING

JACK

PILE IN
SECTIONS

PILED UNDERPINNING

Figure 2.36   Underpinning

former are generally more resistant than the latter and heartwood (the inner part of a trunk) more resistant than sapwood (the outer part of a trunk).

Wood can be damp for two main reasons. If it is freshly felled, particularly in the summer, it will contain sap, a condition referred to as 'green'. Older wood, as used in a building, may have been 'seasoned' (i.e. dried out before use) but can all too easily pick up moisture from defects such as a leaking roof. In either case it is the amount of moisture in the wood which will affect its resistance to attack. Surveyors can measure the moisture content of wood and there are recommended levels as follows:

| Application | Moisture content as percentage of dry weight |
| --- | --- |
| Internal joinery close to source of heat | 8% |
| Internal joinery in continuously heated room | 10–12% |
| Internal joinery with intermittent heating | 15% |
| Rafters and upper floor joists | 15% |
| Ground floor joists | 18% |
| Maximum to avoid fungal decay | 20% |
| Air seasoned timber in UK | 17–23% |
| Kiln dried timber in UK | 0–17% |

These figures are approximations but they are a guide as to how to avoid excessive shrinkage and to combat decay. One has to remember that timber may well have a moisture content over 20 per cent when delivered by a supplier or even more if it has been left uncovered in wet weather. Used in construction, in an ill-ventilated location perhaps, such timber will provide ample food for the agents of decay. The significant figure is 20 per cent; if the moisture content can be held below this, rot cannot develop and beetles will generally lose interest and pass on.

## 2.8.2 Dry rot

Softwood with a moisture content of between 30 and 40 per cent and in a warm location (up to about 26° Celsius) will attract dry rot (*Serpula lacrymans*). This is a particularly destructive

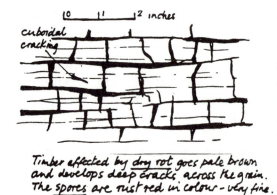

Timber affected by dry rot goes pale brown and develops deep cracks across the grain. The spores are rust red in colour - very fine.

Figure 2.37   Dry rot affected timber

fungus which thrives behind panelling, in flat roofs or under floorboards in houses of any age. Not only does it destroy timber but it can penetrate masonry walls, and spreads either by discharging airborne spores or through threadlike tentacles (hyphae). These tentacles carry moisture with them (which accounts for the Latin *lacrymans* meaning 'tears') and may spread outwards for 1 or 2 metres.

Dry rot destroys the fibres in the wood and results in the characteristic cuboidal crack-ing seen in Figure 2.37. This pattern may start to show on the painted surface of woodwork which will then sound hollow when tapped.

Dry rot has a distinctive 'mushroomy' smell which can give it away and the spores can be detected by people with asthma or hay fever. Dogs can certainly be trained to detect it and specialist firms are now using 'rot hounds' with astonishing success.

When dry rot is discovered, a specialist firm should be called in to deal with it. A general builder is unlikely to have the technical know-how or equipment to be certain of a good job. What the owner or builder can do is to investigate the source of moisture, clear ventilation paths (blocked air bricks in suspended floors, for instance), arrange for damp proofing treatment and do whatever else that may be necessary to dry out the building. Depending on the conditions this may take many months, during which time the situation should be carefully monitored.

A recommended procedure for dealing with dry rot is as follows:

1   Find cause of damp and deal with it.
2   Provide better ventilation.

3   Assess severity of attack. Further action may not be needed if the attack is mild and full drying out is possible.
4   Cut out timber which is structurally damaged and destroy it carefully.
5   Vacuum thoroughly over wide area and dispose of dust in sealed containers.
6   Trace hyphae and expose other affected areas.
7   Sterilize walls with a blow lamp (extreme care needed) and treat with an approved fungicide.
8   Replace all timber which has been cut out with treated timber (see Part 2.8.5).
9   Replace boards, linings etc. in such a way that they can be removed in order to inspect the repair regularly until drying out has been completed.

### 2.8.3 Wet rots

Like dry rot, wet rots thrive when the moisture content of the timber is between 30 and 40 per cent. Wet rot (Coniophora puteana) and cellar fungus (Coniophora cerebella) are not as serious as dry rot because they do not carry their moisture with them and so are more localized. Cellar fungus requires a moisture content of 25 per cent before it will attack wood. It produces masses of white hyphae. Wet rot fungus produces threadlike brown or black hyphae and its main victims are external door and window frames, cills, facias, and boarding where they are exposed to rain and have been lacking perhaps in essential maintenance.

In treating wet rot affected woodwork, the affected material can be cut out and replaced with treated timber but severe attacks may justify whole replacement of a component such as a door or window frame. The advice of a skilled carpenter/joiner used to repair work should be sought.

As with dry rot, it is important to remove the source of moisture to eradicate wet rot. Drying out the affected area as quickly as possible and ensuring that painting, pointing,

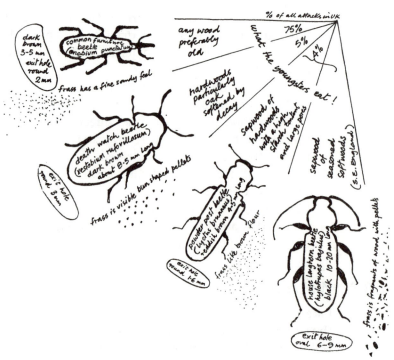

*Figure 2.38   Wood-attacking beetles*

sealants, flashings and so on are functioning correctly will reduce the risk of further outbreaks.

### 2.8.4   Beetles

The characters shown in Figure 2.38 are the main wood-attacking beetles in UK buildings. They can be identified by the following:

- the beetle or larva (if available),
- the flight or exit holes,
- the frass (excreta).

In addition to the four shown, wood boring weevils can be found attacking very damp and decayed wood. They are very small.

Beetles are not so dependent on moisture as fungi although dampness does encourage an attack. It is the grubs or larvae which cause the damage and they generally prefer the sapwood to the heartwood. Timbers are not normally destroyed by the larvae but a gradual loss of strength is evident.

Bad infestations may weaken structural timbers so that they must be replaced. Mild infestations though are frequently treated with an insecticide by brush or spray. Some insecticides are dangerous and must only be used by a trained operator. There is much to be said for a soft approach to a beetle problem. Much so-called 'woodworm' in old buildings is no longer active and has only a marginal effect on the structural performance of the timber. Even brushing away the affected material, 'defrassing' as it is called, is no longer deemed necessary but, as in the case of fungal decay, continuing to observe a suspected timber is important, particularly looking out for clean boreholes and fresh little piles of frass. Look out for the beetles when they emerge in the period June to August (death watch a little earlier, say April to June) but the larvae are almost impossible to see without cutting open the timber.

Death watch beetles are so called because of the tapping sound they make during mating, i.e. in early summer. They can be a severe problem in historic buildings where they attack the heartwood in oak frames. This often takes place at high levels, in a church roof for example, which makes inspection very difficult. In such cases fumigation with smoke bombs has been found effective when carried out annually for, say, 10 years. The cost of scaffolding is thus saved. Insects emerging to mate and lay eggs (in April and May) are poisoned, and fall to the floor where they can be collected and counted – a progress check on how effective the treatment is.

## 2.8.5 Timber treatment

The term 'treated timber' has come to mean timber which has been treated with a liquid preservative either by brushing, spraying, immersion or pressure impregnation.

Dry wood, that is, wood having a moisture content within the limits discussed in Part 2.8.1 above, has an indefinite life and for an exposed application such as internal joinery it needs no protection against decay. In such cases, painting has more to do with decoration than anything else. Where wood is likely to be intermittently damp, for example external cladding or frames, then good quality heartwood should be used and treatment is recommended.

When there is a risk of timber becoming permanently damp, possibly through a fault, and in places where replacement would be expensive such as floors, the use of treated timber is strongly recommended.

The choice of treatment depends on circumstances and, in the case of pressure impregnation, its availability. Most softwood structural sizes are available in treated form but in small works of repair and alterations treatment on site may be acceptable. Jobs requiring timber in quantity such as house building or extensions should use impregnated timber, particularly where an order can be placed in good time.

Thoroughness is the key to good timber treatment, so that whatever the method used it is important to treat cut ends, notches and so on which are made on site. Contractors will not use treated timber unless asked to do so because it costs more than untreated. There is an exception to this in the case of work in those parts of Berkshire and Surrey which are the habitat of the house longhorn beetle and where the building regulations require that roof timbers must be treated. Elsewhere it appears to be a decision for the owner based on the costs and potential benefits of treatment.

Some comparative prices of pressure impregnated timber are given in Table 2.1.

***Table 2.1 Comparative prices for untreated and treated timber***

| Section | Untreated | Treated |
| --- | --- | --- |
| 50 × 100 | 85 | 94 |
| 50 × 150 | 119 | 136 |
| 50 × 200 | 190 | 216 |

nb.prices are pence per metre length for softwood timber as at November 1997.

## 2.9   Listed buildings

Since the Town and Country Planning Act of 1968 it has been necessary to obtain listed building consent for any work on a listed building. This includes everything from repairs to extensions. It is the partial demolition and the consequent loss of historic fabric which is the main problem and the local planning authority may well need to consult advisory groups such as the Georgian Group before coming to a decision. Also at risk is the building's 'character' which an alteration or conversion may affect and this has been a contentious issue for many years.

In fact, concern for the welfare of historic buildings has been growing for about 150 years. In the mid-nineteenth century, John Ruskin objected to the activities of Victorian architects who were 'restoring' historic buildings rather than conserving them. Later, in 1877, William Morris founded the Society for the Protection of Ancient Buildings (SPAB) to be dubbed 'antiscrape' for its opposition to the then prevalent practice of removing historic surfaces before restoration. Gradually over the years the basic conservation philosophy which finds favour today emerged and restoration techniques declined. But it was not until the end of World War II that historic buildings came under statutory protection. Now in 1998 there are probably over 450 000 listed buildings in Britain. PPG15 gave the following breakdown in 1994:

Grade I      Considered to be of exceptional interest: 9000 buildings

Grade II*     Important buildings of more than special interest: 18 000 buildings

Grade II     Buildings of special interest which warrant every effort to preserve them: 413 000 buildings

(*Source: Planning Policy Guidance No. 15: Planning and the Historic Environment* (1994) HMSO)

Most listed buildings were built up to about 1840 but some built later by eminent architects are included and since 1993 a few modern buildings have been listed or are being considered for listing by the Department of National Heritage. Not only buildings are listed: walls, individual stones, cranes, bridges, signposts and so on have been considered of sufficient historical interest to justify listing. The full lists for a particular area can be inspected at the offices of the local planning authority and make interesting reading.

English Heritage (EH) was set up by the government in 1984 as its official advisor on heritage (the equivalent in Wales is Cadw, in Scotland Scottish Heritage). Apart from its role as an owner of historic property, EH is called in by local authorities to advise on conservation issues, provides money for worthy projects and publishes a great deal of useful literature. Its influence on conservation practice has been considerable although not always without controversy as to what should be conserved and the methods to be used.

Where alterations to a listed building are proposed, an application for listed building consent must be made at the same time as applying for planning permission. This will attract the attention of the conservation officer of the local planning authority who will judge the proposal for its likely effect on the character of the building and its historic integrity. Its structural condition will be taken into account as well as its intended use, particularly if the building is dilapidated when extensive repair and replacement work may be more acceptable than if it were not in such a condition.

Some of the guiding principles to be considered are as follows:

- The building must be surveyed, measured and a full record made before any works are started.
- Accurate details of all work carried out must be kept.
- Alterations and repairs should have minimal effect on the historic fabric – as little as possible to be done.
- No work should make the existing fabric any less accessible than it is at the outset.

- Where possible, work should be reversible – capable of being altered later without damaging the fabric.
- Materials used should be compatible with the original ones.
- The replacement of inferior materials used in previous repairs to be a matter for agreement with the conservation officer. They may be deemed to be part of the history.
- Repairs should be 'honest' and no attempt made to disguise them. What is new should be noticeable without being obtrusive.

There are no hard and fast rules about the design of extensions to listed buildings but most conservation offers will look for something 'in keeping' of appropriate scale, form and size, and having no adverse effect on the character of the original. Owners and practitioners who have experience of listed buildings will be familiar with some of the constraints imposed by the local planning authority when granting listed building consent. These can include:

- a strong preference for the use of traditional materials such as handmade clay bricks, wooden windows, lime mortars, natural stone and clay tiles;
- a strong aversion to the use of 'modern' materials such as machine-made bricks, PVCu windows, cement mortars, artificial stone and concrete tiles;
- replacing defective materials 'like with like' such as cast iron gutters with cast iron, bath stone with bath stone, water reed thatch with water reed, and not using more readily available substitutes;
- ensuring that extensions and alterations do not adversely affect the structural integrity of the original building (not allowing indiscriminate removal of structural fabric, for example);
- ensuring that extensions and alterations do not adversely affect the environmental conditions of the original building (realizing the importance of ventilation, for example).

# Further reading

Building Research Establishment (1989) *Simple measuring and monitoring of movement in low-rise buildings (Digest no 343)*. HMSO.

Building Research Establishment (1992) *Remedial Wood Preservatives: Use them safely (Digest no 371)*. HMSO.

Burberry, P. (1992) *Environment & Services* (7th Ed). Longman.

Cunnington, P. (1991) *Care for old houses* (2nd Ed). A & C Black.

Curd, E. F. and Howard, C. A. (1996) *Introduction to Building Services*. MacMillan.

Glover, P. (1996) *Building Surveys* (3rd Ed). Butterworth-Heinemann.

Haverstock, H. (1993) *Easibrief*. Morgan Grampian.

MOHLG (1961) *Homes for Today and Tomorrow*. HMSO (The Parker Morris Report).

Noy, E. (1990) *Building Surveys and Reports*. BSP.

Ranson, W. H. (1993) *Building Failures: diagnosis and avoidance* (2nd Ed). Spon.

Reid, E. (1984) *Understanding Buildings*. Longman.

Richardson, B. (1991) *Defects and Deterioration in Buildings*. Spon.

Richardson, C. (1985) *A J Guide to Structural Surveys*. Arch Press.

# 3

# Valuation and costs

In Chapter 1 the point was made that in the early stages of a project, design and cost are often regarded as equally important, particularly by the client who will also be aware of the potential value of the property on completion. In this chapter these points are developed further. Some practical approaches to valuation are described, followed by alternative ways of preparing cost plans or budgets. Finally some of the more important factors influencing costs in design are considered.

## 3.1 Valuation

Valuation is not an exact science. Basically it is estimating the amount of money a prospective purchaser is prepared to pay and that depends on the state of scarcity or surplus in the supply of the product. Dealing costs must be added for sellers or subtracted for buyers.

Taking an example from the antiques trade, an object might be valued at £1000 when selling but it would have to be insured for, say, £1500 which would cover the cost of a replacement, including the dealer's costs in terms of finance, expertise, staff and other overheads.

In housing the situation is more complicated in some ways because most people buy with the assistance of a mortgage so it is not merely their personal valuation which matters but also that of the mortgagers, i.e. usually a bank or building society. However, it is less complicated in that most costs are broadly standardized. Estate agents' fees are usually percentages of purchase or sales prices and are available, although in some cases can be negotiable. Solicitors' fees can be ascertained. Banks and building societies quote their arrangement fees if asked. Government taxes on transactions are standardized. Most house valuations use one or more of the following methods.

### 3.1.1 By comparison

The valuation of domestic property is usually a matter of comparison. Similar properties in the locality which have been sold recently are a useful guide if the price at which they are sold, as opposed to the asking price, can be ascertained from the estate agent. Adjustments will be necessary: an addition for an attractive front garden perhaps, or a subtraction for a poor state of repair externally. Knowledge of the internal state of a property is important but not so easy to obtain. Further adjustments should be made for the state of the housing market and perhaps the time of year. By these means a fair valuation can be reached.

Owners normally find it easier to be objective about the flaws in a neighbour's house than in their own and the complaint from estate agents is that people usually over-value their own property, but surely that is human nature.

### 3.1.2 Land and construction costs

New and individual properties are quite difficult to value. It is a matter of adding the cost of the site to the cost of design and construction and adding a percentage for selling costs. The result is very close to the development costs we refer to in Section 3.2.

The main problem here is that the cost of good building land rises steadily due to its scarcity and the policies which control the release of land suitable for development. Construction costs are very variable. There are regional variations and a wide range of possibilities depending upon the quality of materials and finishes, the form or shape of the building and the duration of the contract. These matters are more fully discussed in Section 3.3 below.

With the aid of published cost data, price books and so on it is possible to predict building costs in the detail sufficient for budgetary planning. Some of the methods used are described in Section 3.2 below.

The problem may arise that if a large and expensive house is built in a 'down at heel' location it will not be valued at its cost. This is one reason why developers prefer large sites: they have the opportunity to dictate the quality of an area apart from the opportunity such sites provide to effect economies of scale.

The 'nail-biting' experience for people who buy a plot with the intention of building their own home (the self-builders) is that there is a period when the value of the plot diminishes as far as an objective valuation is concerned, because the scheme which has started may have limited appeal to anyone else; it is not their brainchild and the building is not yet capable of use. Moreover, the value of the site as a piece of land may well have depreciated in the meantime.

### 3.1.3 Rental values

Another method of valuation is by calculation of yield. An estimate is made of the prospective rental value per annum. This will take account of such variables as location, accessibility, accommodation, amenities and so on. From this a deduction is made for maintenance and management fees and a percentage for vacancies between lettings. The resultant sum is multiplied by a factor of from 8 to 10. This then is the estimated purchase price.

For example: *a two bedroom apartment*

| | |
|---|---:|
| Estimate rental value | £500 per month |
| Estimate rental value | £6000 per annum |
| Deduct for management | £1000 |
| Deduct for maintenance (incl. VAT) | £500 |
| Deduct for vacancy (1 month) | £500 |
| Total deductions | £2000 |
| Leaving | £4000 |
| At a factor of 10 this would value the apartment at | £40 000 |
| At a factor of 8 this would value the apartment at | £32 000 |

In practice, with a new property maintenance should be low for a year or two and a lettings-only service could be assumed, thereby avoiding management costs. Vacancies may be as low as two weeks in the year so that costs could be halved and a higher valuation with rising rents could be anticipated. However, there is something to be said for being rather pessimistic in these expectations.

### 3.1.4 The alternatives

The options available to a prospective purchaser wishing to value a property can be summarised as follows:

1   Is the purchase an investment for yield?
    If so, use the rental value method (Part 3.1.3).
2   Is the objective to break even and wait for capital appreciation?
    If so, use the comparison method (Part 3.1.1) and add something for desirability.
3   Is the owner intending to live in the property?
    If so, use a combination of all three methods described.

### 3.1.5 Conclusions

Three broad conclusions emerge from this discussion on valuation:

1   The number of dwellings available to rent may vary with the state of the market. When house prices are low the rental market will increase because investment gives a better return on capital. When house prices are high and rising, owners tend to let property only while waiting for capital appreciation or to ensure that a dwelling does not deteriorate through being vacant, i.e. it is kept warm, ventilated and free of intruders.
2   Valuations that depend on comparisons will vary dramatically with fluctuations in the housing market but valuations based on land and construction costs are less volatile. This method, though, is almost impossible to use with older properties because the costs of rebuilding the structure are very different from those of newly constructed

buildings. Foundations are deeper, insulation standards constantly increasing, heating and ventilation more elaborate and so on.

3   Valuations inevitably depend on a number of subjective factors including:

(a)   a preference for a particular building style which in the case of an old building may be associated with the local vernacular;

(b)   a preference for certain materials: brick is popular, concrete is not, and so on;

(c)   the perceived environmental and amenity value of an area;

(d)   the perceived social standing of a neighbourhood when compared with others in the same town or region.

## 3.2   Predicting costs

Owners usually need some assurance early on that the building works they have in mind are affordable. Sometimes it is a case of determining what can be done within a predetermined budget; at other times it is a question of knowing what is wanted and estimating its likely cost. In either case the owner needs to be aware that costs fall into two main categories:

● *Initial or capital costs* – including the cost of building, i.e. the 'contract sum' paid to the contractor and other 'development costs' such as the price paid for the property, design fees, interest on capital, solicitor's fees and so on. Most owners will expect the design team and contractor to establish and control the cost of building but not become involved with other development costs. The price of a plot, for instance, will be vitally important to the prospective purchaser but of little interest to the architect, unless of course they are one and the same person.

● *Costs-in-use* – the costs of ownership or running the building such as heating, lighting, cleaning, maintenance, and so on. These are not applicable where the property is to be sold except in the sense that a building which is economical to run should be a more marketable product. Many decisions made at the briefing and design stage will affect costs-in-use. The design of windows, for example, must include some consideration of the method and therefore the cost of cleaning them. The selection of a floor finish will undoubtedly depend to some extent on the level of maintenance required. These are questions which must be addressed well before design and specification are complete.

The only indication of a market price for a building job is provided when tenders are received but, as we shall see later in Chapter 9, this can only happen after a considerable period of time. In fact, this is so late in the process that should the price be unacceptable a good deal of abortive work may have been done, particularly by the design team. In any case the whole design process can be something of a 'shot in the dark' without the benefit of working to some kind of cost guidelines. It is not surprising that 'cost planning' has become an important part of the preliminary design process and that quantity surveyors in particular will offer it as part of their service. Cost planning uses techniques of approximate estimating but is usually done quickly and with very limited design information such as sketch plans only. It should not be confused with the later processes of tendering and estimating where more complete information is available, although a well-prepared cost plan can be a useful basis for later tendering (see Section 9.3).

There are several ways of producing a cost plan for a building job. Some are described below and their relevance to minor works and alterations discussed. In all cases it is the contractor's tender price for the building work which is being predicted, not the costs to the contractor which are his affair. Neither are we dealing here with overall development costs or costs-in-use which, as we have seen, are matters for separate consideration by the owner.

### 3.2.1 Costs per unit floor area

The Building Cost Information Service (BCIS) which is provided by the Royal Institution of Chartered Surveyors (RICS) publishes quarterly statistics of current tender prices for a wide range of building types. They are presented in the standard form of £ per square metre of gross floor area, i.e. the total floor area of a building measured between the outside walls and including voids and circulation spaces. Typical examples of average prices for the third quarter of 1996 were as follows:

Private detached houses $£729/m^2$

Two-storey estate housing $£391/m^2$

Four-storey apartments $£511/m^2$

Factories generally $£328/m^2$

Offices generally $£724/m^2$

These are directly applicable and a convenient guide to new building costs since all that is required is an estimate of the floor area of the proposed building. As with all such statistics, however, it is necessary to forecast trends far enough forward to cover the anticipated timing of the work. The BCIS helps with this too by publishing tender price indices (TPI), which can be used in much the same way as the Retail Price Index (RPI) is used to predict future trends in the cost of goods and services. For example:

- In 1985, when the tender price index was 100, a job cost £10 000.
- In June 1998, when the tender price index is expected to be 146, the same job would cost

$$\frac{10\ 000 \times 146}{100} \quad = \quad £14\ 600.$$

Another adjustment which is required is to allow for regional differences. In February 1997, for instance, average tender prices in the London area were 18 per cent higher than in North East England, 20 per cent higher than in Wales and 51 per cent higher than in Northern Ireland. The considerable differences in market conditions across the country which these figures illustrate are not always appreciated.

This method of forecasting costs based on gross floor area cannot be usefully applied to alteration work with its widely variable nature, but where a relatively large extension amounts

99

to a 'tacked on' piece of new build it will have some relevance provided that allowance is made for the following:

- the work involved at the join between existing and new, often difficult to assess and quantify until partial demolition is under way;
- the fact that an extension which includes a relatively high proportion of services and fittings (typically a bathroom and kitchen) will have a higher cost/m² than one that does not;
- the likely additional load on services created by the extra space and the possible need therefore to upgrade a boiler, electrical intake, drainage and so on.

## 3.2.2 *Approximate quantities*

This method requires drawings to have been sufficiently well advanced so that measurements can be taken from them. The drawings need not show the detail required for precise measurement but should show clearly the amount, location and the nature of the work involved together with an outline specification. Even then the exercise requires considerable skill in reading the drawings and visualizing all the processes involved with the construction. Whether this should be done by a quantity surveyor or a technician depends rather on the size and complexity of the job.

Having compiled a list of quantities or 'measured items', these are priced using 'all in' rates which may be obtained from builders' price books such as Spon's or Wessex. These are the rates which allow not only for the work described but also for the cost of related items not separately measured. A typical item from a bill of approximate quantities would read as follows:

33F: One brick wall in common bricks and 1:1:6 gauged
      mortar: 20 m²

The equivalent item from the 1997 edition of Spon's would read:

Common bricks in gauged mortar (1:1:6) one brick thick: £36.31

From this it can be calculated that the price for the item would be:

$$20 \times £36.31 = £726.20$$

### 3.2.3 Schedule of works

This is a method which does not require drawings or specifications to have been prepared but does require the owner and/or the design team to have established what needs to be done. A typical domestic conversion or extension will often comprise about 20 to 30 operations or activities and these may be used not only for early planning but for programming and cost control later on. Having prepared the schedule in draft form, the ideal arrangement is for the contractor to be brought in, to 'walk the job' with the owner and to discuss the most effective methods of proceeding. Having agreed the schedule it is then ready for pricing and will become a tender document. If the contract is to be negotiated, the contractor who assists with the schedule will be the one who prices it and will no doubt be involved with further discussion on alternative prices before the final tender is agreed. (See Chapter 9.)

A typical priced schedule of works for a small alteration job is shown in Table 3.1.

The advantages of this method are:

- The contractor's expertise with regard to construction methods, materials, economies, programming and so on can be taken into account early in the design process.
- With the contractor's co-operation, the cost plan can be produced quickly and well in advance of design information.
- The priced schedule becomes the basis of the contractor's tender and a contract document which facilitates cost control as the works proceed.
- It is a relatively simple process which does not rely on the availability of costing books or systems.

The main disadvantage of the method is that it is a form of negotiation and does not therefore go well with competitive tendering. In the latter situation all competitors should be treated equally, which is an obvious difficulty if one of them has had an influence on design decisions. This can be avoided, of course, by producing a schedule of works without the involvement of a contractor but the probability is that more variations will occur later on. The advantages and disadvantages of negotiated contracts are discussed more fully in Chapter 9.

## Table 3.1 A priced schedule of works

14th February, 1997.

Ref: Mill Street, Oxford. ESTIMATE

| Job no. | Work | Value £ |
|---|---|---|
| 1 | Carefully remove chimney stack from kitchen and bedroom above. | 1,460.00 |
| 2 | Remove partitions to WC and external store and remove WC and cap. | 420.00 |
| 3 | Block in WC and store doors and render to match existing. | 346.00 |
| 4 | Remove existing kitchen window and form opening for new in kitchen gable wall. | 140.00 |
| 5 | Supply and fit Magnet Statesman hardwood windows to openings DG ref. 210C and 2N10C. | 780.00 |
| 6 | Cut back part of floor under stairs to form WC compartment and make good. | 360.00 |
| 7 | PC sum to lay new drain to WC with branch to SVP, including connection to existing and forming new manhole. | 550.00 |
| 8 | Inject silicone DPC with 30-year guarantee. | 500.00 |
| 9 | Remove all wallpaper. | 600.00 |
| 10 | Replaster as DPC spec. including all further areas. | 2,100.00 |
| 11 | Lay new solid floor to kitchen to current regs. | 810.00 |
| 12 | Lay quarry tiles, labour cost £420 included. | 912.00 |
| 13 | Remove and strip doors and including cupboards. | 250.00 |
| 14 | PC sum for new ironmongery. | 250.00 |
| 15 | Strip matchboarding in hall and at side of stairs in kitchen. | 350.00 |
| | External decorations: | |
| 16 | To previously painted joinery: | 620.00 |
| 17 | Weathershield to render to rear extension: | 450.00 |
| | Subtotal of estimate | 10,898.00 |
| | VAT @ 17½%: | 1,907.15 |
| | Total | 12,805.15 |

## 3.2.4 Elemental method

Part of the service provided by the BCIS is to provide historic cost data in the form of cost analyses of completed projects. Each cost analysis follows a standard format based on the building's elements, which helps when making comparisons and forecasting future cost trends. By selecting suitable examples from the large number of cost analyses available, it is possible to analyse these and to arrive at a breakdown of likely elemental costs for a particular building type. As we have seen, though, it is necessary to take account of regional trends and to apply indices for future projections. Differences in specification must also be considered.

A typical cost analysis for a small housing scheme in Wolverhampton completed in 1994 is shown in Table 3.2.

There is an abundance of cost analyses available in the BCIS system, particularly for various types of housing, but for years they have been based on new build. Only in 1995 did BCIS begin to introduce cost analyses for rehabilitation/conversions and inevitably it will be some time before large enough samples can be used for analysis. Even then, unlike new building projects, works to existing buildings differ so greatly that matching examples to a future project will be very difficult. What is useful is knowing how much certain elements such as heat, electricity, staircases and windows actually cost. What may also be gleaned from a study of cost analyses for a particular building type are trends in the way costs are apportioned. Taking two-storey mixed housing as an example, the average percentage costs are as shown in Table 3.3.

For new build this may be interesting as it stands but if we assume that in a conversion job the substructure and superstructure of the building are not affected then the cost of the remainder should lie within about 50 per cent of what the equivalent new build would cost. This is an over-simplification of just one way in which BCIS data can be used to predict costs.

To be fully effective this method should be used by someone who is familiar with the system and has access to it, typically a quantity surveyor or surveying technician. Otherwise it can only be cost-effective for new build projects of a reasonable size.

## 3.3  Factors influencing the cost of building

The BCIS statistics show that for any building type the variations in tender prices (i.e. cost to the employer) are very wide. Sheltered housing up to 1996, for example, varied from £251 to £1023 per square metre with an average or mean of £493

## Table 3.2 A typical cost analysis

| Element | | Preliminaries shown separately | | | Preliminaries spread | | |
|---|---|---|---|---|---|---|---|
| | | Total cost | Cost per m² | Element unit quantity | Element unit rate | Total cost | Cost per m² |
| 1 | Substructure | 60,869 | 79.15 | 587 m² | 103.70 | 67,869 | 88.26 |
| 2A | Frame | – | | | | | |
| 2B | Upper floors | 5,167 | 6.72 | 182 m² | 28.39 | 5,761 | 7.49 |
| 2C | Roof | 40,795 | 53.05 | 818 m² | 49.87 | 45,487 | 59.15 |
| 2D | Stairs | 2,274 | 2.96 | 5 No | 454.80 | 2,536 | 3.30 |
| 2E | External walls | 46,878 | 60.96 | 740 m² | 63.35 | 52,269 | 67.97 |
| 2F | Windows and external doors | 58,386 | 75.92 | | | 65,101 | 84.66 |
| 2G | Internal walls and partitions | included in | 2E | | | | |
| 2H | Internal doors | included in | 2F | | | | |
| 2 | Superstructure | 153,500 | 199.61 | | | 171,154 | 222.57 |
| 3A | Wall finishes | 46,139 | 60.00 | | | 51,445 | 66.90 |
| 3B | Floor finishes | included in | 3A | | | | |
| 3C | Ceiling finishes | included in | 3A | | | | |
| 3 | Internal finishes | 46,139 | 60.00 | | | 51,445 | 66.90 |
| 4 | Fittings | 9,078 | 11.80 | | | 10,122 | 13.16 |
| 5A | Sanitary appliances | 7,256 | 9.44 | 39 No | 186.05 | 8,090 | 10.52 |
| 5B | Services equipment | – | | | | | |
| 5C | Disposal installations | 965 | 1.25 | | | 1,076 | 1.40 |
| 5D | Water installations | 6,010 | 7.82 | | | 6,701 | 8.71 |
| 5E | Heat source | – | | | | | |
| 5F | Space heating and air treatment | 30,988 | 40.30 | | | 34,552 | 44.93 |
| 5G | Ventilating systems | – | | | | | |
| 5H | Electrical installations | included in | 5F | | | | |
| 5I | Gas installations | included in | 5D | | | | |
| 5J | Lift and conveyor installations | – | | | | | |
| 5K | Protective installations | – | | | | | |
| 5L | Communications installations | 4,500 | 5.85 | | | 5,017 | 6.52 |
| 5M | Special installations | – | | | | | |
| 5N | Builder's work in connection | 7,080 | 9.21 | | | 7,894 | 10.27 |
| 5O | Builder's profit and attendance | – | | | | | |
| 5 | Services | 56,799 | 73.86 | | | 63,330 | 82.35 |
| | Building sub-total | 326,385 | 424.43 | | | 363,920 | 473.24 |
| 6A | Site works | 54,207 | 70.49 | | | 60,441 | 78.60 |
| 6B | Drainage | 27,169 | 35.33 | | | 30,294 | 39.39 |
| 6C | External services | 45,183 | 58.76 | | | 50,379 | 65.51 |
| 6D | Minor building works | – | | | | | |
| 6 | External works | 126,559 | 164.58 | | | 141,114 | 183.50 |
| 7 | Preliminaries | 52,090 | 67.74 | | | – | |
| | Total (less Contingencies) | 505,034 | 656.74 | | | 505,034 | 656.74 |
| 8 | Contingencies | 6,850 | 8.91 | | | 6,850 | 8.91 |
| | Contract sum | 511,884 | 665.65 | | | 511,884 | 665.65 |

**Table 3.3 Average percentage costs for two-storey houses**

|  | Average % | Range |
|---|---|---|
| Substructure | 9 | 6–13 |
| Superstructure | 34 | 25–44 |
| Internal finishes | 9 | 6–13 |
| Fittings | 2 | 1–3 |
| Services | 13 | 9–17 |
| External works | 23 | 21–33 |
| Preliminaries | 9 | 3–17 |
| Contingencies | 1 | 0–4 |

per square metre. As we have seen, some of this can be attributed to regional differences but there are other factors affecting tender prices and the most important of these are briefly discussed here.

## 3.3.1 Length of contract

Some building jobs will cost more than others of equivalent size because they take longer to build. This is because certain costs such as insurance, site supervision, plant hire and temporary works such as scaffolding are time related. When a contract does not finish on time some or all of these costs will increase, an expense to be borne by either the employer or contractor depending on the provisions in the contract between them.

In practice, these factors are itemized in specifications and bills of quantities as 'preliminaries'. In large or medium-sized contracts they are usually priced separately from measured work but in small jobs they are normally included in the contractor's prices for measured work. In the analysis shown in Table 3.3 the preliminaries amounted on average to 9 per cent of the total price but the range was from 3 to 17 per cent.

## 3.3.2 Specification

This is perhaps the simplest of these factors and one that the owner and design team must discuss at the design and cost planning stage. As we saw in Chapter 1, materials and components are selected to fulfil precise functions or to provide a certain level of performance which stems from what the

105

designer perceives to be the owner's requirements. However, in practice there is almost always the constraint of costs, particularly where the owner is looking for the lowest possible price for the job. In some elements of the work there will be little choice in the materials to be used – using plasterboard for the ceilings or treated softwood for floor joists, for example. However, in other elements there are often alternative specifications with wide-ranging cost implications, such as:

- Internal blockwork – plastered and painted or just painted?
- Kitchen floor – ceramic tiles or vinyl tiles?
- Staircase – hardwood or softwood?
- Roofing – concrete tiles or clay tiles?
- Heating – gas fired only or partly solar?

These questions must be addressed having due regard not only for initial costs but also costs-in-use arising from maintenance requirements, energy usage and the expectation of necessary repairs.

As we shall see in Chapter 9, it is the contract documents which must clearly indicate the required standards of materials and workmanship. Failure to spell out precisely what is required will almost certainly lead to variations and possibly conflicts later on.

### 3.3.3 *Form of building*

Simplicity in the form of a building or extension usually leads to an economic design. Simple forms, such as a cube, lead to uncomplicated detailing, enclose the internal space with the minimum amount of wall and simplify foundation and roof construction. This last point is important because the most cost-efficient building is one that uses materials in a structurally efficient way. There are optimum spans for standard size joists and rafters, for example, and waste can be avoided by using materials such as plasterboard in standard sizes to avoid cutting. See Figure 3.1.

Flat roofs are the exception to the general rule that simplicity of form and compactness are desirable aims. Apart from being out of fashion on aesthetic grounds, their relatively inexpensive initial cost tends to be outweighed by high maintenance and repair costs whereas the alternative pitched roof of slates or tiles

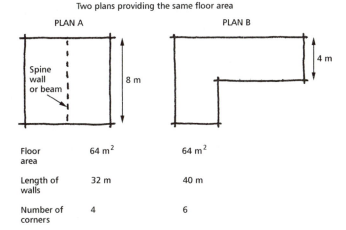

Two plans providing the same floor area

| | PLAN A | PLAN B |
|---|---|---|
| Floor area | 64 m² | 64 m² |
| Length of walls | 32 m | 40 m |
| Number of corners | 4 | 6 |

On these figures A would be less expensive to build than B but A requires longer spans than B and to use 50 × 200 floor joists would need a load bearing spine wall or beam pushing up the cost of A. Another point is that 50 × 200 joists at 400 mm centres will span 4.2 m with domestic loads but a more economic span would be 4.0 m because 4.2 m is a standard length for such timber and a minimum of 100 mm bearing must be allowed at each end. For similar reasons 3.4 m is an economic span for a standard joist length of 3.6 m although in this case only joists of 175 mm depth would be required.

*Figure 3.1   Comparative costs of building forms*

is not only visually pleasing but carries relatively low costs-in-use. It should be said that much of the bad press inflicted on flat roofs can be attributed to poor workmanship, particularly in failing to provide a sufficient fall so that the roof drains well and in failing to use roofing felt correctly. Modern flat roofs are less prone to failure but experience and skill are still required if they are to perform well.

## 3.3.4   Nature of work

The nature of building work and the levels of skills required will have a bearing on costs. Even owners who have only ever had repairs done to their properties will appreciate this. There is a tendency among people unfamiliar with building work to assume that defects in elements like drainage would be more difficult and expensive to deal with than familiar ones like rotting windows. In practice, costs relate less to physical effort and more to the level of skills required, manufacturing costs and the quality of materials used. So renewing a window could well be relatively expensive compared to renewing a drain, even if a mechanical digger had to be used.

The cost of renewing a roof again depends on the trade skills and materials involved but in older houses some of the cost can be set against the opportunity to provide better ventilation, insulation and fire protection, possibly even incorporating roof windows or solar panels at the same time. The greater life expectancy of the building is also a bonus.

Repairs and alterations to listed buildings demand special attention. Apart from the constraints imposed by the local planning authority, the techniques and materials used should be of a particular quality and this will invariably lead to above-average costs. To some extent this is not justified, particularly where a contractor new to conservation work inflates his prices to cover his ignorance. On the other hand, some traditional processes such as oak framing, leadwork and thatch are undeniably more costly than their modern equivalent.

### 3.3.5 Costs-in-use

Earlier we discussed the need to balance predicted capital costs and costs-in-use at an early stage in the project. There are a number of issues here. The concept of maintenance-free buildings is an attractive one but difficult to achieve given the environmental conditions and servicing arrangements that most buildings endure. On the other hand, a reduction in maintenance costs is often feasible provided the appropriate decision is made at the design stage. For example, in the case of windows the following options can be considered:

| | |
|---|---|
| Avoid need for external painting | Use PVCu, aluminium or hardwood frames (in the latter case from sustainable sources) |
| Avoid need for external maintenance | Design so that windows can be cleaned from the inside |
| Avoid need for external repairs | Detail so that window glass can be replaced from the inside |

It should be noted, of course, that the relevance of this approach will be governed by the height of the building. If

**Table 3.4   Cost, life and maintenance comparisons**

| Material | Initial cost | Life | Maintenance cost |
| --- | --- | --- | --- |
| Mild steel | Low | Moderate | High |
| Stainless steel | High | Long | Low |
| Brickwork | Low | Long | Low |
| Weatherboarding | Moderate | Moderate | Moderate |
| Carpet tiles | High | Moderate | Low |
| Ceramic tiles | Moderate | Long | High |
| Hardwood | High | Long | Moderate |
| Softwood (untreated) | Low | Moderate | High |
| Clay roof tiles | High | Long | Low |
| Felt roofing | Moderate | Short | Moderate |

scaffolding can be avoided then a considerable saving can be made in both capital and maintenance costs.

In general, and assuming competent design and workmanship, higher capital costs can lead to longer life and lower costs-in-use, particularly where repairs and replacements are concerned. Some generalized comparisons are given in Table 3.4.

# Further reading

Ashworth, A. (1994) *Cost Studies of Buildings* (2nd Ed). Longman.

Building Cost Information Service (updated regularly) *Manuals*. RICS.

Cartlidge, D.P. and Mehrtens, I.N. (1982) *Practical Cost Planning*. Hutchinson.

Davis, Langdon and Everest, Eds (yearly) *Spon's Architects' and Builders' Price Book*. Spon.

Johnson, V.B., Ed. (1996) *Laxton's Building Price Book* (169th Ed). Laxton.

Seeley, I.H. (1996) *Building Economics* (4th Ed). Macmillan.

# Professional assistance

In this chapter the roles and responsibilities of some of the professionals involved with building projects are discussed. Professional institutions are listed and notes provided on typical conditions of engagement.

## 4.1 Who does what?

A wide range of professional skills is involved in buying, altering and letting or selling houses. These are encompassed by the following professions:

| | |
|---|---|
| Architect | Planner |
| Accountant | Quantity surveyor |
| Building surveyor | Services engineer |
| Estate agent | Solicitor |
| Interior designer | Structural engineer |
| Landscape architect | Valuer |

It is assumed that readers may be interested in performing some of these roles themselves, but such action may not be cost-effective, possible or enjoyable so it is necessary to consider each of these professional roles and assess what or who is needed for each project. They are taken here in alphabetical order. The addresses of appropriate professional bodies are appended to each part. Note that these building-related roles should not be seen as the totality of any particular professional's activities.

### 4.1.1 Architect

**Main roles**

- Helping clients to clarify objectives and discussing alternative ways of meeting them

- Advising on choice of property, need for regulatory approval and selection of contractors, sub-contractors etc.
- Preparing draft schemes for client's approval
- Submitting plans for outline or full planning approval and dealing with appeals if necessary
- Preparing detailed drawings for building regulation approval
- Preparing working drawings and specifications for construction
- Drawing up contract and tender documents
- Acting as a planning supervisor under the CDM regulations
- Advising on selection of tenders
- Issuing instruction as the work proceeds, particularly when difficulties arise
- Supervising the work on site
- Authorising payment of contractor's accounts

### General note

Architecture is the widest of all the professional disciplines related to building and an architect's training includes more than a passing reference to quantity surveying, structural engineering, planning, urban design, interior and landscape design. Many architects add a related discipline, such as landscape architecture or town planning, to their initial training. Architectural courses include design projects which vary considerably in type and size, typically from small-scale domestic works to major housing developments and opera houses.

After training an architect will gravitate towards the scale and type of work in which he or she is most interested but many practices welcome small-scale work as an opportunity for training junior employees as well as a design challenge in its own right. An architect may not be a member of the Royal Institute of British Architects (RIBA) or the Architects and Surveyors Institute (ASI) but must register with the Architects' Registration Board (ARB, formerly ARCUK) in order to use the title Architect. In Scotland the professional body is the Royal Incorporation of Architects in Scotland (RIAS), in Wales the Royal Society of Architects in Wales (RSAW) and in Ulster the Royal Society of Ulster Architects (RSUA). Other practitioners may be architectural technicians and be qualified members of the British Institute of Architectural Technologists (BIAT).

The architect's forte is his or her ability to design, that is, to interpret the client's requirements and transpose those into buildable forms. No one else in the design team receives such a thorough training in design. Architectural technicians are often extremely capable of carrying out design procedures and sometimes more capable than their professional colleagues in dealing with the technicalities of design, but a skill in creative design may have eluded them.

Contractors who offer design as a service to their customers often appoint, or enter into partnership with, an architect for that purpose. Once frowned upon, this would appear to be a growing practice.

In 1995 the Construction (Design and Management) Regulations 1994 (CDM) came into force. These require architects as designers to consider safety in their design work. In addition the architect may be appointed as a 'planning supervisor' and have particular duties under the Regulations. The implications of this are discussed in Section 5.4.

## Contact

RIBA
66 Portland Place
London W1N 4AD
Tel: 0171 580 5533

ASI
15 St Mary Street
Chippenham
Wilts SN15 3JN
Tel: 01249 444505

ARB
73 Hallam Street
London W1N 6EE
Tel: 0171 580 5861

BIAT
397 City Road
London EC1V 1NE
Tel: 0171 278 2206

RIAS
15 Rutland Square
Edinburgh EH1 2BE
Tel: 0131 229 7545

RSUA
2 Mount Charles
Belfast BT7 1NZ
Tel: 01232 323760

RSAW
Midland Bank Chambers
75a Llandennis Road
Cardiff CF2 6EE
Tel: 01222 762215

## 4.1.2 Accountant

### Main roles

- Advising clients on business planning and taxation
- Preparing accounts for submission to tax authorities, either HM Customs & Excise for Value Added Tax (VAT) or HM Inland Revenue for income and capital gains taxes
- Preparing annual balance sheets and profit and loss accounts for companies as required under the Companies Acts
- Preparing financial projections, such as cash flow, profit and loss, for submission to funding bodies, banks etc.
- Acting as an auditor for companies and organisations.

### General note

If an individual sells a house which has been his or her residence, it is generally free of taxation, unless it is done so frequently that it obviously becomes a business or a form of self-employment. An accountant will advise on the timing of operations and whether to aim for capital gains, which has a higher tax-free threshold, or income tax, which allows a range of costs to be set against income. Preparing accounts for small works with the aid of a guide may well be a useful activity for most householders but it is certainly advisable to discuss matters at the outset with an accountant, particularly because a professional would be aware of recent or pending legislation. Professional accountants are usually members of the Institute of Chartered Accountants in England and Wales (ICAEW) which has a membership of well over 100 000 individuals and 19 000 firms. Scotland and Ireland have similar bodies.

### Contact

ICAEW
PO Box 433
Chartered Accountants' Hall
Moorgate Place
London EC2P 2BJ
Tel: 0171 920 8100

Institute of Chartered Accountants of Scotland
27 Queen Street
Edinburgh EH2 1LA
Tel: 0131 225 5673

Institute of Chartered Accountants in Ireland
11 Donegall Square South
Belfast BT11 5JE
Tel: 01232 221600

## 4.1.3 Building surveyor

### Main roles

- Producing measured drawings of buildings
- Inspecting and reporting on the condition of a building
- Identifying significant defects and essential repairs
- Preparing schedules, drawings and specifications for works of repair, maintenance and adaptation
- Acting as a planning supervisor under the CDM regulations
- Making an assessment of market value before and after the work
- Making an assessment of reinstatement value for insurance purposes
- Advising on particular problems associated with historic buildings
- Acting as a contract administrator for a building project

### General note

Qualified building surveyors are members of the Royal Institution of Chartered Surveyors (RICS) which oversees their training and sets out their professional code of conduct. Mortgage companies employ building surveyors to survey and advise on a property's eligibility for a loan but this advice is not available to a prospective purchaser even though he or she will be charged the survey fee. This is one reason why a prospective purchaser is recommended to commission an independent survey and report, preferably using the format of the RICS/ISVA Home Buyers' Survey and Valuation Report which we referred to earlier in Section 2.2.

Many estate agents are managed by qualified building surveyors and staffed by people at various stages of their training, but estate agents are usually agents of the vendor of a property so this is another reason why a purchaser should employ an independent surveyor unless they are satisfied that they can perform that role themselves in a fully competent fashion.

The building surveyor's training is a broad one and provides a capability of dealing with a range of diverse activities particularly related to the appraisal of existing properties. An experienced building surveyor will have a good grasp of the principles of building structures, materials, environmental factors, contract procedures and design. He or she should be well equipped to deal with the complexities of altering, converting, extending or rehabilitating buildings and nowadays it is not unusual to find a surveyor who has a good understanding of the special needs of historic buildings.

### Contact

RICS
12 Great George Street
London SW1 3AD
Tel: 0171 222 7000

## 4.1.4 Estate agent

### Main roles

- Acting usually as an agent for sellers of property
- Acting sometimes on commission to search for and buy property for a client (in which case the buyer pays the agent's fee)

- Assisting in obtaining mortgage and insurance if required
- As letting agents, advising on suitability of property for letting, finding tenants and managing property thereafter

### General note

Estate agents are responsible for producing 'particulars' of the properties they deal with. Particulars are often described as 'thought to be materially correct ... accuracy not guaranteed'. Nevertheless errors are sometimes made and an agent may be prosecuted and fined for making false claims. Although an estate agent's immediate interest is to sell a property and thus may appear to encourage a vendor to accept a low price for a quick sale, he or she is also motivated by the need for continuity of work. Today's seller may be tomorrow's buyer. Satisfied customers and personal recommendations are important so agents will tend to balance their concern for both buyer and seller.

### Further points to note

- A good estate agent is scrupulous over the handling of bids and offers for properties but a higher bid coupled with some delay is less likely to succeed than a lower one made for early completion.
- An agent should take reasonable steps to ensure that a prospective buyer has the means to proceed with a purchase and that a seller can clear a mortgage, if any, by selling.
- Establishing good relations with an estate agent is useful for a developer who, by so doing, will learn of potential development opportunities well before they become widely available through newspaper advertising which is subject to weekly deadlines and so on.
- Frequent transactions allow agents to offer developers a lower rate of commission.
- Estate agents will also advise on the likely selling price of a property after upgrading and fairly standard alterations but should not be expected to envisage dramatic or unusual changes.
- Because they are in frequent contact with local builders and developers they may be aware of rough guide prices for standard works.

Estate agents may be members of the National Association of Estate Agents (NAEA) but this is not mandatory and many will be chartered surveyors or valuers.

**Contact**

NAEA
Arbon House
21 Jury Street
Warwick CV34 4EH
Tel: 01926 496800

## 4.1.5   Interior designer

**Main roles**

- Preparing design schemes, or alternatives, for the interior of a building
- Designing permanent fittings such as shelves, cupboards, screens and so on
- Selecting items and sources of supply for fabrics, furniture, light fittings etc.

**General note**

Interior design may well fall within the remit of an architect or at least be the subject of close co-operation with an architect so that overall design and detailed design can be co-ordinated. Furthermore, since artificial lighting is so often a part of interior design, it is often necessary to bring the electrical engineer into the early discussions. However, for the client, interior design is a very personal matter and one which most householders thoroughly enjoy. In this case, design assistance is often given by major furnishing stores, paid for indirectly by commission on the goods they supply. Often people with an eye for decorative style are only too happy to make suggestions, and books and magazines proliferate. Where professional help is required, advice may be sought from Interior Decorators and Designers Associated Ltd (IDDA) or the Chartered Society of Designers (CSD). A qualified designer may provide drawings with samples of fabrics, floor finishes and wall colours. Judging the impact of a large area of colour can be difficult but professional

117

experience helps. Also DIY tips, such as painting the inside of a cardboard box to assess the effect of reflective colour, can save time.

### Contact

IDDA
1–4 Chelsea Harbour Design Centre
Lots Road
London SW10 0XE
Tel: 0171 349 0800

CSD
32–38 Saffron Hill
London EC1N 8FH
Tel: 0171 831 9777

## 4.1.6 Landscape architect

### Main roles

- Surveying a garden to assess potential and problems
- Advising on trees (age, health, eventual size etc.) particularly those close to buildings
- Designing garden planting
- Designing 'hard' landscaping – paths, retaining wall, pergolas, ponds and so on
- Producing drawings when required by the local planning authority (often as a condition of receiving planning permission)

### General note

The potential of a garden depends on the type of soil, aspect, drainage, climate, shape and degree of shelter, all of which need to be appraised at the start.

The appearance of a garden is a matter of personal taste – restrained whites and greys against a dark green background, vivid colour tumbling over stone walls, lofty trees set in level lawns, terraces on sloping ground and so on. A landscape architect can suggest ways and the relative ease of achieving particular effects, taking due account through experience of local environmental conditions. A common problem for lay people in planting a new garden is to choose trees and shrubs which rapidly outgrow their intended space. Familiar though most people are with some plants, help is often required in sorting out the many varieties now available in garden centres and specialist nurseries. Garden centres may offer advice on planting schemes, using their stock of course, but it is clearly preferable to have an independent advisor who has visited the garden in question. Qualified landscape architects are members of the Landscape Institute (LI) which has a membership of about 3900 including 260 or so overseas. The strong connection of landscape with horticulture and gardening of course usually means that there is no shortage of local 'garden designers', some offering a 'complete garden service' and so on. These are worth investigating preferably by seeking out and scrutinizing their work after a few seasons of weathering.

### Contact

LI
6/7 Barnard Mews
London SW11 1QU
Tel: 0171 738 9166

## 4.1.7  Planner

Qualified planners are members of the Royal Town Planning Institute (RTPI) and may act in one of two main roles.

### As a consultant

- Advising on procedures associated with town planning applications, legislation etc.
- Commenting on the feasibility of alternative development proposals
- Preparing layout plans and documents, submitting these to the local planning authority (LPA)
- Negotiating with the LPA on matters of potential difficulty (non-compliance with a local plan, for example)

- Dealing with appeal procedures

### As an LPA planning officer

- Responding to enquiries from potential applicants
- Dealing with planning applications under the general heading of 'development control'
- Drawing up development plans (district, local, area etc.) which form the basis of local planning policy
- Advising the local planning committees on the acceptability of planning applications

In neither of these cases does a planner produce detailed plans of buildings. In fact, planning today is much more about policy making than plan drawing. That is not to say that the LPA will not look at detail when assessing an application. On the contrary, planners are often critical of architects' design proposals and this can lead to conflict.

The planning profession started with the Garden City movements of the nineteenth century and achieved independent status after 1945 to cope with the problems left by wartime bombing, a rapid growth in population and an even more rapid growth in traffic. Thus planners were primarily concerned with the planning of neighbourhoods and the relationship between schools, hospitals, workplaces, residential areas and the transport system. This was 'land use' planning.

The past few decades have seen a number of changes; for example, a greater concern for the social environment of planning has attracted graduates from such disciplines as human geography and the social and environmental sciences; architects with an interest in urban problems have moved into the field of urban design in preference to planning; the planning graduate of today will be concerned with a much wider range of environmental issues than in the past; historic conservation has become a major consideration and has led to the appointment of conservation officers to the planning departments of local authorities. Apart from all these concerns the planner must still deal with the day-to-day problems of development control.

**Contact**

> RTPI
> 26 Portland Place
> London W1N 4BE
> Tel: 0171 636 9107

## 4.1.8 Quantity surveyor

### Main roles

- Producing 'cost plans' (budget statements) and advising on financial feasibility of design proposals
- Calculating the quantities of work, in terms of labour, plant and materials, in a building job (a process known as measurement)
- Compiling bills of quantities which can be 'priced' by those tendering for the job
- Checking and advising on tenders achieved
- Valuing the work as it proceeds (usually on a monthly basis) so that the contractor is paid as each stage of the work is completed
- Negotiating with the contractor on the pricing of variations and on intermediate and final accounts
- Advising on contractual matters generally
- Acting as a contract administrator for a building project
- Acting as a planning supervisor under the CDM regulations

### General note

Basing an estimate on the cost of materials alone, which may be a lay person's approach, is a mistake since it overlooks the relatively high costs of labour and equipment. For example, the cost of a wall with openings may involve no saving in brick-work over one without because of the labour costs in forming the openings. Having available bills of quantities ensures that everything that contributes to the cost of work is taken into account and ensures that where competitive tenders are sought the process is equitable and thorough. Without bills contractors will calculate their own quantities and this can sometimes lead to much variation and omission in the prices submitted. The acceptance by the client of the lowest tender produced under such circumstances will almost certainly lead to contractual problems later (see Section 9.3 on tendering and estimating).

Professional quantity surveyors are members of the Royal Institution of Chartered Surveyors (RICS) and are engaged either in consultancy practices or as employees of large client or contractor's organizations. In either case they are expert in the production of bills of quantities but this is a relatively expensive process which cannot be justified for a small job. In this case, the production of a cost plan and assistance with cost control as the work proceeds are the most useful services the quantity surveyor can provide. Alternatively his or her expertise in contractual matters and financial control may be best used in the role of a contract administrator, project manager or design team leader even though, in the latter case, the architectural profession might not agree. Since under the CDM regulations the preparation of specifications and bills of quantities are seen as 'design', the QS now has a duty to consider safety in the preparation of these and must liaise closely with other design team members on matters of safety and risk assessment.

Contractors frequently employ quantity surveyors either as permanent staff or as agents to deal with contractual matters such as valuations, cost control, claims as a result of variations, final accounts and so on. Working with his or her opposite number, the 'private' QS, the contractor's quantity surveyor can have a strong influence on the running of a contract. When financial difficulties crop up it can be reassuring to have a QS 'on your side'.

### Contact

> RICS
> 12 Great George Street
> London SW1P 3AD
> Tel: 0171 222 7000

## 4.1.9 Services engineer

### Main roles

- Advising, at an early design stage, on alternative schemes for:

  > Plumbing and sanitation
  > Heating and ventilation
  > Electrical power

Lighting

Vacuum cleaning

- Producing drawings and specifications for the above
- Advising on the selection of tenders
- Supervising installations

## General note

Professional services engineers, who will be members of the Chartered Institute of Building Services Engineers (CIBSE), are normally only employed on medium to large contracts. Even when they are employed, and certainly when they are not, it is common practice for services design work to be undertaken by a specialist sub-contractor. The cost of design is then included in the price that the sub-contractor quotes for the work. In cases like this, it is not unusual for tenders to be invited based on performance specifications only, so that the detailed design work is only carried out by the successful tenderer. In simple domestic jobs the performance standards are relatively easy to establish after discussions between the engineer and the client.

Such matters as the air temperature in rooms and therefore the level of heating required, the amount of ventilation and so on are well established in standard codes and regulations. Electric lighting and power, though, offer enormous scope for variation in the owner's requirements, and lighting particularly is often closely related to architectural or interior design. Even the disposition and number of power sockets are often crucial to the successful function of many interiors. For this reason it is the electrical engineer or sub-contractor who is the most likely person to have to liaise with the architect or designer so that electrical layout plans can be produced for the client's approval. Both at the design stage and later as the work proceeds, the service installations must be integrated with the other building processes. Drawings showing the position of fittings, cable routes, pipe runs, switches, control panels and so on are not always necessary on small jobs but they can be useful as working drawings on site and in the case of historic buildings they will enable the owner and architect to assess the likelihood of damage to the building's fabric and to take avoiding action.

It should be noted that services installations must be carried out by competent operators who are fully conversant with current safe practice and legislation. Gas installations, for example, must be done by a firm registered with the Council for the Registration of Gas Installers (CORGI). Services design and installation must comply with CDM regulations.

*Contact*

CIBSE
Delta House
222 Balham High Road
London SW12 9BS
Tel: 0181 675 5211

CORGI
4 Elmwood
Chineham Business Park
Crockford Lane
Basingstoke
Hants RG24 6WG
Tel: 01256 707060

## 4.1.10   Solicitor

*Main roles*

- Advising on all legal matters
- Acting on behalf of one of the parties to a contract in the event of a dispute with the other
- In respect of disputes, negotiating settlements in advance of the more formal processes of arbitration or court proceedings
- Conveyancing, i.e. the legal processes involved with buying and selling property, on behalf of either the buyer or the seller

Buying involves:

- searches with local authority, i.e. ascertaining the local authority's plans for an area such as road widening, conservation areas etc.;
- obtaining contract (deeds) and studying the implications of any covenants;
- advising the client of the need to insure against certain risks,

an old right of way being exercised for example;
- handling money on behalf of the client and mortgager.

A large legal practice may also arrange insurance cover for the property to be purchased for a short period until the client is able to arrange their own insurance. A solicitor will also draw up or advise on any tenancy agreements and on any subsequent problems which may occur in respect of a tenancy.

Selling involves:

- drawing up and sending out a contract to the purchaser or purchaser's solicitor;
- answering purchaser's queries which may require consultation with the vendor;
- supplying relevant documents such as past planning approvals, listed building consent, building regulation certificates, etc.;
- handling money on behalf of client, estate agent and mortgager.

### General note

Buying is the most onerous of the conveyancing solicitor's roles, partly because of the legal principle of *caveat emptor* or 'buyer beware'. Not all solicitors specialize in conveyancing so a potential purchaser could back up a solicitor's work with private enquiries, keen observation and common sense. Not all solicitors are skilled at reading plans and may seek professional help in this respect. A solicitor is in the fortunate position in relation to house transactions in that he or she handles the money and can deduct his or her fees at source without delay. This contrasts with much of the construction industry which is plagued by late payment.

Solicitors should belong to The Law Society of England and Wales (The Law Society) which has a membership of 66 000, or The Law Society of Scotland or Northern Ireland.

### Contact

The Law Society
Law Society's Hall
113 Chancery Lane
London WC2A 1PL
Tel: 0171 242 1222

The Law Society of Scotland
26 Drumsheugh Gardens
Edinburgh EH3 7YR
Tel: 0131 226 7411

The Law Society of Northern Ireland
98 Victoria Street
Belfast BT1 3JZ
Tel: 01232 231614

## 4.1.11   Structural engineer

### Main roles

- Advising on all matters pertaining to structural stability, including the condition of an existing structure and the likely effect of altering it
- Preparing drawings, specifications and calculations to satisfy local authority building control with particular regard to:

  > Foundations and ground conditions
  > Unusually large spans or heights
  > Roof frameworks

- Checking site practice and quality with regard to structural items such as reinforced concrete, steelwork and heavy timber engineering
- Acting as a planning supervisor for engineering projects as required under the CDM regulations

### General note

In normal circumstances and with traditional forms of construction, an architect would not need to invoke the assistance of a structural engineer. The use of proprietary components such as steel lintels, beams and roof trusses means that calculations, if required, can be obtained from manufacturers who must at the same time of course guarantee their products. Nevertheless, altering buildings produces frequent non-standard situations where the expertise of a structural engineer will be required. As we saw in Section 2.3, he or she can have an important role to play when the survey of a building reveals structural inadequacies. The structural engineer tends to be a specialist, unlike architects, surveyors and civil engineers whose work is more generalized and who are likely to pass on roles with which they have become less familiar.

Professional structural engineers are members of the Institution of Structural Engineers (IStructE) and most work in consultant practices. Many started their professional life as civil engineers and are also members of the Institute of Civil Engineers (ICE). Local building inspectors often require that structural design work should only be carried out by qualified professional engineers even though the original building form may be designed in the first case by an architect. Ideally, of course, architectural design and structural design should be a matter of partnership.

### Contact

IStructE
11 Upper Belgrave Street
London SW1X 8BH
Tel: 0171 235 4535

ICE
1–7 Great George Street
London SW1P 3AA
Tel: 0171 222 7722

## 4.1.12  Valuer

### Main role

- To assess the value of property
    for purchase
    for replacement
    for investment

### General note

Most of the valuer's work in relation to houses is incorporated in other roles, e.g. building surveyor, estate agent, mortgage company. An independent professional valuation may be helpful in the case of a dispute with an insurance company over a problem such as subsidence, fire damage and so on. The more usual role of a valuer for a householder is in the assessment of the value of the contents, possibly before an auction, to decide on marketing strategy. In this context it is advisable to keep photographs and information on size, date, artist, manufacturer, author etc. of all items of value and to update the list regularly.

A professional valuer will be a member of the Incorporated Society of Valuers and Auctioneers (ISVA) and most probably be working in partnership with estate agents and surveyors to provide a broad range of services to building owners and prospective owners. The Home Buyer's Survey and Valuation package referred to earlier is a result of the close relationship between the ISVA and the RICS.

### Contact

ISVA
3 Cadogan Gate
London SW1X 0AS
Tel: 0171 235 2282

## 4.2 Obtaining professional assistance

The professional institutions named above will advise potential clients on the selection of suitable members, particularly those who are available locally and who are prepared to undertake small to medium-sized projects. Usually a short list of names and addresses is provided and it is then up to the client to decide how many people to contact for preliminary discussions. It may be that the first person contacted, an architect for example, will recommend another person, a quantity surveyor perhaps, with whom he or she has worked in the past. The value of assembling a small team of people who have already practised together successfully should not be overlooked. Many of the larger design organizations are essentially group practices in which architects, engineers, planners and so on collaborate in providing a comprehensive service to their clients. This may not be appropriate for most medium to small projects but good informal teamwork can bring benefits to works of any scale.

Should the approach through a professional institution not bring about the desired result, the following should be considered:

- *Personal recommendation* – ask friends and colleagues who have had work done for their comments on the performance of the consultants involved.
- *Observation* – look out for building work in the locality. Sometimes there are professional sign boards at the site entrance but a good-looking site should be a reflection of a good working team (or at least an efficient contractor). Enquiries can be fruitful.
- *Public notices* – local authorities are obliged to publish lists of planning applications, often in their local daily or weekly newspaper. These are useful guides as to which agents are active in the area and what kind of projects they are handling.
- *Local officials* – building control officers of the local authority may recommend specialists such as a structural engineer who has experience of dealing with particular forms of construction (medieval timber frames, for example) or with certain local materials (earth walling, for example).
- *Directories* – perusing the *Yellow Pages* or other local directories, or newspaper advertisements, will reveal lists of practitioners which can be useful, but only as a first step. Ultimately there is a need for the other forms of investigation described above.

## 4.3   Conditions of engagement and related matters

A person engaging the services of an architect, engineer or surveyor is entitled to expect a skilful and competent service. Membership of a professional institution ensures that a member is aware of what is expected of him or her, even though, regrettably, the service provided may not always reach the standards required by the client. Generally by appointing a 'professional' an adequate level of competence is provided. The institutions would argue, with some justification, that if a 'non-professional' is appointed instead then competence is a matter of chance. The difficulty is that many very capable practitioners are competent, even well qualified, but have not actually become members of their respective professional bodies. This may be from choice, lack of funds or, as in many cases, from a lack of experience since professional practice is a prerequisite of professional membership. To take one example, a recent graduate with a degree in architecture will need to work in an architect's office for two years and complete a diploma course of two years before becoming eligible for RIBA membership. In the meantime he or she is a capable designer who needs experience. Clients would be well advised to talk to such a person, consult his or her referees, study his or her portfolio of work and consider the contribution that such a person could make to a project where design flair is needed. The overall responsibility for the architectural service could still be placed in the hands of a chartered architect or even perhaps a chartered surveyor. Similar situations can arise where graduate surveyors and engineers are capable practitioners but have yet to achieve professional status. In some respects they may have more advanced skills than their more senior colleagues and one should perhaps be aware of their potential even in relatively small projects.

### 4.3.1  Architects' services and fees

The RIBA divides the architect's services specific to building projects into 11 work stages known as the Schedule of Services or Plan of Work. These are:

A   Inception
B   Feasibility
C   Outline proposals
D   Scheme design
E   Detail design
F   Production information
G   Bills of quantities
H   Tender action
J   Project planning
K   Operations on site
L   Completion

Each stage comprises a number of activities in the form of a checklist from which the client and architect can select by mutual agreement the tasks to be performed. It is not unusual for the client to require only a partial service such as design work and planning approval (up to and including Stage D), design and production information for Building Regulations approval (up to and including Stage F) or everything up to receipt of tenders (Stage H). In the latter case the client would be making others responsible for site supervision.

The Schedule of Services serves another useful purpose in that it provides a framework for the architect's fees. Traditionally architects' fees were based on fairly rigid scales but they are now much more likely to be negotiated and they vary considerably from firm to firm and job to job. If the fee is calculated as a percentage of the total construction cost, which is common practice, they would normally be from 6 to 9 per cent for new housing and from 9 to 15 per cent for housing conversions or alterations. Generally speaking the larger the contract the lower the percentage fee. Repetitive work tends to command a low fee but the added complexity of works to existing buildings has always justified a relatively high fee. Where a partial service has been agreed or where payment of the fees is to be in instalments, the following percentages of the total fee may be applicable:

| Work stages | Fee |
|---|---|
| A and B | Usually a lump sum based on the amount of time spent |
| C | 15% |
| D | 20% |
| E | 20% } of the full service fee |
| F and G | 20% |
| H, J, K, L | 25% |

Thus for designing a scheme and obtaining planning permission (up to and including Stage D) the architect would be paid 35 per cent of the fee payable for the full service, i.e. Stages C through to L.

## 4.3.2 Conditions of engagement

Clearly it makes good sense for both the client and his professional agent to have a clear understanding of the terms under which they are to work together. The professional institutions have model 'conditions of engagement' which clarify the scope of the service being offered, establish certain procedures and define the limits of liability. These are quite lengthy documents and since the 'small print' of such things is often overlooked it can be helpful if at least the essential conditions are referred to in any letter of appointment. As an illustration only of the nature and content of such documents, the RIBA Conditions of Engagement (CE/95) are paraphrased below. This is a simplification. The reader should read the whole document to appreciate its finer points. Note that both the architect and the client are required to be 'reasonable' at times, giving some scope, of course, for differences of interpretation and argument in due course.

### RIBA Conditions of Engagement for the Appointment of an Architect (CE/95)

- *Governing law/interpretation*
  The laws of England and Wales/Scotland/Northern Ireland to apply (as appropriate).
- *Architect's obligations*
  To exercise 'reasonable skill and care'.
  To act on behalf of the Client.
  To keep Client informed of progress, changes to design or expenditure.
- *Client's obligations*
  To provide all information 'reasonably necessary'.
  To advise on priorities re budget and project timetable.
  To make decisions on time.
  To accept that the Architect does not warrant the work of others.
- *Assignment and sub-contracting*
  Neither party to assign without the consent of the other.
  Architect not to sub-contract services without consent of the Client.
- *Payment*
  To be in accordance with an agreed schedule of fees and expenses.
  Percentage fees to be based on total construction cost as forecast or achieved.
  Fees may be varied according to services provided.
  Architect is entitled to additional fees for extra work.
  Payment to be by instalments on submission of Architect's account.

Interest to be payable on late fees.

VAT may be payable.

- *Suspension, resumption and termination*

  Either party may suspend giving reasonable notice in writing but in the Architect's case only if the Client is in default of payment of fees.

  Either party may terminate the appointment with reasonable notice in writing.

- *Copyright*

  Copyright of all documents and drawings and work executed is the Architect's property.

- *Dispute resolution*

  Disputes to be referred by either party to an arbitrator to be agreed or someone appointed by the President of the RIBA (in Scotland the Dean of the Faculty of Advocates).

It should be noted that where an architect is appointed as a planning supervisor (CDM) there are additional services which are described in Section 5.5. Inevitably the breadth and complex nature of the architect's service has led over the years to the development of a broad and complex set of conditions. In the emerging climate of competition and litigation in the 1990s it would be an unwise practitioner who failed to heed the advice of his professional institution in these matters. The RICS also publish Standard Conditions of Engagement which can be used for purposes similar to those described above for Architects. Small jobs such as house surveys, though, may not justify the production of multi-paged documents. In these cases a carefully worded but brief set of conditions, covering the essential points, can be quickly drafted and agreed without fuss with the client. One such document is produced below. Note that the surveyor's essential obligation is to 'use all reasonable diligence, skill and care to be expected of a Chartered Surveyor' as we saw in Section 2.2.

### Conditions of engagement for structural/building surveys

1  BS will use all reasonable diligence, skill and care to be expected of a Chartered Surveyor in inspecting and preparing a Report of his findings.

2  In carrying out the Survey, BS will inspect all such parts of the property as are visible from within the property,

but will not lift floor coverings, move heavy furniture, lift floorboards, or carry out any investigation which might be destructive of the substance or decorations of the property.

3　The outside of the property will be inspected from the ground level and from a ladder not exceeding three metres in length, used from the ground floor only. All such inspection will be carried out from within the curtilage of the site and from adjacent public highways only. Outbuildings, paths, drives and boundaries will only be briefly commented upon, unless specifically requested. Swimming pools and associated structures, tennis courts and other such constructions, will not be commented upon.

4　The services of the property comprising electrical supplies and wiring, gas, soil drainage, water and central heating will be inspected visually, but no tests made unless the Client specifically instructs the use of specialists to undertake such work and pays the actual costs incurred in employing such specialists.

5　The Client understands that BS is a visitor to the property which belongs to another party and that the party has the final say as to what tests and inspections may be made. In the event that such party prevents what BS considers to be normal inspections, this will be the subject of comment in the survey report.

6　BS shall be entitled to assume that no noxious, dangerous or deleterious matter has been used in the construction of the property or incorporated subsequently, unless such matters are revealed by visual inspection, without further analysis. BS shall be under no obligation to make inspection of the wall cavity, if any, but shall report on the apparent condition of the property, including reference to any suspected wall tie failure. BS shall not inspect the foundations, but shall make recommendations for such inspection if there is evidence to suggest that such an inspection is necessary.

7　In making any valuation of the property, BS shall be entitled to assume:

(a)　that the tenure of the property is as described by the Client and that there are no unusual restrictions, conditions or encumbrances which have not already been brought to his attention;

(b) that the property is not subject to any matter which may be revealed by the Local Searches, Enquiries or Statutory Notices and that the erection of the property and its subsequent use is not contrary to any Town Planning or other similar restriction, and that the use to which the Client intends to put the property is not excluded by any such matters;

(c) that inspection of the parts not inspected would not reveal material defects which would cause BS to alter his opinion as to the value.

8 In giving any assessment as to the likely cost of repair, BS will not have received formal estimates for the work. Any such figure is, therefore, intended as a rough guide to likely costs and not a calculated estimate.

9 In respect of those parts of the property which have not been examined in accordance with these Conditions, the Client shall not be entitled to assume that no defects exist. Reference to some defects in any particular part of the property is not a guarantee that other defects do not exist – minor defects will not be noted individually.

10 The Report of the Survey is confidential to the Client and the Client's professional advisors. The Client shall be entitled to disclose such sections of the Report as required to the Vendor of any property which is the subject of the Survey. BS accepts no responsibility to any other person reading the Report and any such person who relies on the Report does so entirely at their own risk.

11 Before commencing the Survey, the Client will sign a letter accepting these Conditions of Engagement, together with the fee quoted in the letter accompanying them. Any fees of specialists employed at the Client's request will be payable immediately on receipt of the account.

### 4.4.3 Professional indemnity insurance

Chartered surveyors, incorporated valuers, chartered engineers and architects are advised by their institutions to maintain professional indemnity insurance. This is a complex matter but the intention is to cover the costs arising from claims of negligence up to a certain limit which should be stated in the letter of appointment or 'memorandum of agreement'. Costs may be considerable because they may include not only the cost of putting right a defect but also the 'consequential loss' arising from the disruption caused by putting it right. A loss of production in a factory would be a consequential loss, for example.

Professionals may be sued for negligence under either the law of tort or contractual law and, considering the number of errors it is possible to make in the design, specification or quantification of building work, the risks are high. Very large sums of money are involved and the premiums to be paid are high enough to cause many a headache for the lone professional. Another consideration is the period of liability which, as far as building defects are concerned, is governed by the Latent Damage Act 1986. We shall be discussing this in Section 9.5.2 under defects liability.

Professional indemnity insurance is a subject where the expert advice available within the appropriate professional institution should be sought. It is wise to take out insurance before practising because policies will not normally cover negligence alleged to have occurred in the past.

## Further reading

Aqua Group (1992) *Pre-Contract Practice* (8th Ed). Blackwell.

Aqua Group (1996) C*ontract Administration for the Building Team* (8th Ed). Blackwell.

RIBA (1994) *Architect's Guide to Job Administration under the CDM Regulations*. RIBA.

RIBA (1995) *Condition of Engagement for the Appointment of an Architect* (CE/95). RIBA.

RIBA (1994) *Guidance for Clients on Fees*. RIBA.

Scott, J.J. (1985) *Architectural Practice*. Butterworth.

Walker, A. (1996) *Project Management in Construction* (3rd Ed). Blackwell.

Willis, C.J., Ashworth, A. and Willis, J.A. (1994) *Practice and Procedure for the Quantity Surveyor* (10th Ed). Blackwell.

# 5

# The legal framework

This chapter is concerned firstly with the general legal background and with some long-established constraints on development, and then with the regulations which govern planning and building control, and finally with the Construction (Design and Management) Regulations.

English law involves long-established common law, legal judgements which set a precedent and Acts of Parliament, sometimes followed by Statutory Instruments and Orders produced by government departments. The system has evolved over time and the legal framework is periodically extended as ideas about activities which require regulation change and the powers of the EU increase.

The categories of law which most affect development are civil law and administrative law. Civil law includes contract law, which is considered here because the building process involves a series of contracts between people, and tort, the legal word which implies the right to sue for damages in the event of negligence. Disputes may be taken to court. Courts are arranged in a hierarchy and lower courts are bound by precedents set by higher ones. The amount of money at dispute will establish which court applies, with County Courts limited to £3000 unless both parties agree in advance to a higher sum. High Court cases have to be taken by a barrister.

In the case of planning, a simple and normally inexpensive system has been established which enables appeals against some local authority decisions or actions to be taken to the DoE Inspectorate for determination by an inspector or the Secretary of State, but an appeal against an inspector's decision may be taken on legal grounds to a High Court.

# 5.1 Legal constraints

In the course of development, particularly in alteration works, there are some problem areas which are encountered frequently and involve long-established legal constraints which require further consideration; they are as follows.

## 5.1.1 Right of light

Since the Rights of Light Act 1959 a window which has been in existence for 27 years without being blocked acquires a right of unobstructed light. This could prevent new building close to a window. However, the distance from any new wall and the angle of light is not fixed. A rule of thumb of 45° projected upwards from the lowest window cill is generally used to calculate obstruction. But the best course of action is an informal approach to neighbours, or an injunction to stop work could ensue. The use of the room with the affected window may influence the decision of the court.

## 5.1.2 Party walls

Construction work which affected party walls, i.e. a wall shared by adjoining properties, once came under common law but in London from 1939 it was regulated by the London Party Wall Building Act. From 1 July 1997 similar procedures applied to the rest of the country through the Party Wall Act of 1996.

The Act extends earlier common law rights, regulates procedures and provides a framework for preventing and resolving disputes.

*Rights of developers* include the right to:

● raise, demolish and rebuild, or underpin a party wall
● cut into a party wall to take the bearing for a beam
● insert a damp proof course right through a party wall
● protect two adjoining walls by a flashing from the higher to the lower.

*Obligations for developers* are to:

● give notice to adjoining owners
● reach agreement over the work to be done or appoint surveyors to do so

- take care, provide temporary protection and compensate for any damage
- pay the fees of surveyors involved in the works.

### Giving notice

It is best to discuss plans with neighbours and resolve any problems before serving notice; this should pave the way and pre-empt any disputes. There is no set form of notice but guidance on the information the notice should contain is given in the DoE explanatory booklet on the Party Wall etc. Act 1996. Notice has to be given at least two months in advance of the proposed start of the works. If the recipient does not reply within 14 days, a dispute is deemed to have arisen.

### Reaching agreement

If agreement cannot be reached between neighbours or their surveyors then an 'agreed surveyor', i.e. a suitable but not necessarily a professionally qualified person, should be mutually agreed upon and appointed.

The surveyor(s) will draw up a party wall 'award'. This sets out:

- the work to be carried out
- the condition of the neighbouring house
- access for inspection of the works.

### Costs

The cost of the works and the surveyor's fees are usually entirely at the expense of the developer, unless poor repair of the neighbouring property has necessitated the works or the neighbour asks for extra works to be included for their own benefit.

### Minor works

Items such as replastering the party wall or fixing shelves are normally considered too trivial to involve the Act.

*Note.* The above comments are only a brief résumé of an important Act. For more information consult the DoE explanatory booklet, which includes helpful diagrams, and the additional statutes and publications it recommends.

### 5.1.3 Public rights of way

Public rights of way are noted on the 'definitive map' which is an Ordnance Survey map kept with a written description of the route by the county council. Sometimes the exact line is not clear on the deeds of a property or even on the definitive map, and the written description can help to pinpoint the exact line. It is not legally permissible for a developer to divert a footpath, even if the diversion would provide an improved route. An application for a diversion has to be made to the local authority and they would have to advertise any proposed change of route as a diversion (under Section 257 Town and Country Planning Act 1990). If there are no objections or if they are withdrawn, the local authority can confirm a diversion order, otherwise the order and the objections are referred to the DoE who may approve the order, refuse it or revise it. In the latter case the order has to be advertised again. Sometimes diverting a footpath can create development possibilities but it should be noted that there can be strong opposition to diverting a popular route. However, failure to follow the correct procedures can result in a public footpath going through a private garden, or even in a structure built across a path being demolished. The Countryside Commission published a booklet in 1994 entitled *A Guide to Procedures for Public Path Orders*. This is obtainable free from the local authority.

### 5.1.4 Private rights of way

Private rights of way are noted on the title (ownership) deed for a property. Generally they allow access to and from a property over another property, but sometimes the use has not been exercised for many years and the right may have lapsed. Sometimes an alternative public route has been established which is used instead. It may be difficult to establish who owns a private right of way so it may be necessary to insure against resumed use. As a precaution it may be advisable not to build any permanent structure over even a disused private right of way, at least not at first. Gradually screen it with planting.

### 5.1.5 Covenants

Covenants are noted in the deeds of a property. They are legal restrictions or benefits conferred on a property by a previous owner. These may go back to medieval times, such as the permission granted by John O'Gaunt to various house-holders in Hungerford to fish in the local river or graze a donkey on the common. Covenants may also be more recent, such as a limit on the number of houses or type of use on a piece of land. Recent covenants can sometimes be lifted by negotiation with the owners, particularly if they are an institution such as a college. It is very important to note that planning approval does not alter or supersede a covenant. A house converted to flats, for example, might have to be restored to its former use if this change was prohibited by a covenant, despite planning approval.

## 5.2 Avoiding disputes

Small building works often give rise to dissatisfaction on the part of the client over the standard and the amount of the work done, and on the part of the contractor over payment for the work. Disputes can also arise with neighbours over incidental damage or short-term factors such as dust and noise.

Disputes are costly in terms of time and stress and often in terms of finance, so every step should be taken at the outset to ensure they do not arise. In some projects, use of a standard form of contract should help prevent disputes, because informal contracts are liable to be misinterpreted by either side. See Section 9.5 on forms of contract. It is helpful to inform neighbours about any proposed building activities so that they can make any adjustments they need. If problems do arise the first course of action is for the parties to attempt to reach an agreement. If that fails then mediation with the aid of solicitors may be the solution. Any solicitor can mediate on an informal basis but formal mediation can only be undertaken by one who is a member of the ADR Group (Alternative Dispute Resolution Group) or other qualified mediator. See the booklet and leaflets referred to below.

If a claim is made against one party, the best response for the other party may be to bring a counterclaim as well as a defence; it will depend on the circumstances and a solicitor would be able to advise.

Solicitors will need a full statement of what has occurred and when, and copies of any correspondence. Hopefully some agreement will be reached or the matter dropped in the early stages. A good solicitor with a client's financial interests at heart will recommend avoiding taking an action to court. When substantial sums are involved in a dispute the legal costs are very high even when a case is won, and the case might be lost.

A booklet is produced by the Lord Chancellor's Department on 'Resolving Disputes without Going to Court', December 1995. It is free from county libraries. This sets out three ways of resolving disputes:

- Direct negotiation
- Mediation, which means involving the help of a third party (agreed by both sides)
- Arbitration, which means appointing an arbitrator whose decision is binding

The booklet gives very helpful advice, particularly on how to manage direct negotiation and how mediation works. The options for arbitration and the advantages or otherwise of using this method are quite complex and it might be useful to discuss this first with a solicitor, or the Citizens' Advice Bureau. The booklet lists organizations offering mediation and arbitration services. A series of leaflets on small claims and arbitration is available free from County Courts and libraries.

## 5.3 Planning and development control

This section is about:

> The background to the system
> Development and permitted development
> Listed building consent
> Advantages of permitted development
> Making a planning application
> Notes on planning approvals
> Outline planning applications
> Retrospective planning applications
> Planning appeals

### 5.3.1 The background to the system

National policies have been enacted by Parliament to protect towns and the countryside from indiscriminate development and to aim for a balance between housing, employment and infrastructure. The 1947 Town and Country Planning Act is the basis of postwar planning and the present system of development control, followed by key Acts in 1962, 1968, 1971 and 1990. Statutory Instruments and Development Orders also regulate development. Circulars and Planning Policy Guidance Notes (PPG) are issued by the Department of the Environment (DoE) to modify or update current practice.

At local level the development control system is managed by the Local Planning Authorities (LPA), i.e. by the District or City Councils, London Boroughs, County Councils and the new unitary authorities. They implement both national and local policies.

Structure and local plans to guide development are produced by the County and District Councils respectively and, after a public hearing before an inspector appointed by the DoE, may be modified, are then approved and finally adopted for implementation. Local plans may show the location of conservation areas, sites earmarked for development and streets subject to parking or other restrictions. Design guides which offer suggestions on how development, particularly extensions, can be integrated successfully into the local scene are produced by some LPAs.

### 5.3.2 Development and permitted development

Development is defined as 'the carrying out of building, engineering, mining or other operations in, on, over or under land or the making of any material change in the use of any buildings or other land'. So the term 'development' includes change of use, including changing a house to two or more apartments. Change of use requires planning permission. Planning permission is required for virtually all development, with one or two minor but useful exceptions; these are termed permitted development.

#### Permitted development

The parameters for permitted development are complex and not always clear. They are set out in the Town and Country Planning (General Permitted Development) Order 1995. Sections relevant to small-scale residential development are explained in the DoE booklet *Planning – A Guide for Householders*. It is advisable to write to the local planning authority at the outset; a formal reply that planning permission is not required could help in the eventual sale of the house.

Some important permitted development allowances are:

- *Terraced houses (including end of terrace)* – ten per cent of the volume of the original house or the house as it stood, including extensions, on 1 July 1947 or up to 50 cubic metres, whichever is the greater, and up to a limit of 115 cubic metres can be added without needing to apply for planning permission.
- *Any house in a conservation area, national park, area of outstanding natural beauty or the Broads* is restricted to the same limits. Also any additional building of more than 10 cubic metres anywhere in the garden counts against this allowance.
- *Detached houses (except in a conservation area)* – fifteen per cent of the volume of the original house (see above) or 70 cubic metres, whichever is the greater, can be added as permitted development up to a maximum of 115 cubic metres.
- *Buildings in the garden* such as summer houses etc. may cover up to half the garden (in a conservation area the limit is 10 cubic metres) but within 5 metres of the house (or another building belonging to the house) are considered extensions. Height is restricted to 3 m (4 m to apex). However, they may not be placed in front of the house unless there is at least 20 m to the highway.

There may be restrictions on permitted development which are not always obvious, such as:

- *Narrow plots* may restrict permitted development. For example, the restrictions on extensions within two metres of the boundary preclude a first floor extension for most terraced houses.
- *Conditions on planning approvals and Article 4 directions* may remove some or all rights. They are sometimes imposed on new schemes where the site is fully developed and any new building might be oppressive, and so have to be very carefully considered. Single dwellings have permitted development rights whereas flats and apartments do not.
- *Demolition* – in some circumstances demolition counts as

development, particularly in conservation areas, and planning permission is required.

### 5.3.3 Advantages of permitted development

The main advantages are design freedom, and savings in time, cost and uncertainty.

#### Design freedom

Originality in design is often stifled by the planning process. There is a tendency for both designers and authorities to play safe and repeat standard solutions to ensure an easy passage through the system. Permitted development allows some welcome freedom in design.

#### Time

Small straightforward schemes are usually determined by the area planning officer, maybe after consultation with other members of a planning team, but more complex schemes will go to committee meeting(s). In either case there is a process of consultation with the parish council, neighbours and anyone else who might be affected, which takes time. Further information, modifications and assurances are frequently required, all time consuming.

#### Cost

There is a fee for submitting a planning application in addition to the cost of preparing a scheme for formal presentation. This may only be a small proportion of the cost of building works – and the important consideration is to get the right accommodation – but such costs mount up and leave less for the actual work. Delays also tend to be expensive – prices of materials rise and builders may not be prepared to wait or require extras.

#### Uncertainty

Until a scheme is finally approved there is a long period of uncertainty which can make lifestyle and financial planning difficult, and this can become rather stressful. However reasonable a scheme may seem to a developer, either a next door neighbour, a planning officer or a committee member may have a different opinion and a planning application may not be approved. Then there is the problem of how to proceed. In order to gain approval for a scheme it may be necessary to change the design in a manner unsatisfactory to the applicant, which can mean living with permanent irritation. Or an appeal to the DoE may be considered, again involving time, cost and uncertainty.

### Comment

For these reasons it is advisable to do as much work as possible within the framework of permitted development and to plan the sequence of work carefully to make the most of the allowances. For example:

- Consider constructing any extensions which might be needed before applying for a change of use if a house is likely to be converted into two or more apartments.
- If a house is small and a very large extension is needed, it is wise to consider adding the allowance first and then applying for permission to extend further at a later date, because planning authorities tend to limit the volume of extensions to a proportion of the existing dwelling.

### 5.3.4  Listed building consent

Internal and external works to a listed building require listed building consent if they might affect its character. Some LPAs interpret this as any works to a listed building. There is no retrospective permission in relation to listed buildings so it is important not to take risks and to consult the conservation officer of the LPA at the outset. In addition the general restrictions in relation to conservation areas also apply under the Planning (Listed Buildings and Conservation Areas) Act 1995. Extensions to listed buildings may also require planning permission. See also Section 2.9 on listed buildings.

### 5.3.5  Making a planning application

At the LPA, check whether the property is either listed, in a conservation area, subject to an Article 4 direction or any other restriction. Conservation areas are usually depicted by a thick black outline which sometimes slices through a building and can be confusing. The area planning officer will know but the desk staff may not, so it is important to get an authoritative reply. Inspect the planning register for records of planning applications or appeals on the site or nearby, in order to assess the probable outcome of any new proposals. Collect the relevant planning application forms and any helpful guides.

## Application forms

Some LPAs produce three types of forms:

- Minor works and extensions
- New works, change of use and outline applications
- Listed buildings and conservation areas

Sets of forms are packaged with explanatory notes, certificates of ownership and a schedule of fees. It saves time and errors to complete one and photocopy the required number, with one for a record. When submitting the application a certificate has to be included to confirm either that the applicant owns the site (Cert A) or that the owner has been informed that a planning application is being made (Cert B). Without a certificate and the relevant fee the application will not be processed.

## Drawings

The explanatory notes give the information required on the set(s) of drawings. Applications for full approval generally need to show each elevation and floor plan to 1:50 or 1:100 scale, with the roof shown on the site plan to 1:500 scale. A location plan to 1:2500 is required with the site outlined in red and any other land owned by the applicant outlined in blue.

## Notes on presentation

Show the orientation with a north point; this helps to show whether overshadowing might occur, as well as locating the site on Ordnance Survey maps. Existing buildings may need to be shown to explain how the new work will relate. Diagrams and photographs, perhaps with sketches superimposed, sometimes provide useful extra information. Consider the scale; large buildings to a large scale can look gross. If size is likely to be a controversial factor it is advisable to present the drawings to a small scale and to emphasize the landscape setting. The presentation should be attractive whether it is precise or free in style; it may be pinned up on the committee wall.

### Terminology

Words should be used carefully in a planning context. The word 'office', for example, implies commercial use to planners and if all that is intended is home work the word 'study' is more appropriate and is unlikely to arouse opposition and delay.

### Consultation

It is sometimes advisable to send a covering letter with an application stating that the applicant and/or their architect would be pleased to meet and discuss the application if there are any items at dispute. This should prevent a summary rejection and will allow a period to make modifications.

### Processing the application

An application is acknowledged and then checked and if it is in order, it is numbered and registered. The registration date is important because if no decision is given, i.e. the scheme is not determined, any appeal has to be made within eight months of that date. However, an extension of time may be requested by the LPA. Plans will be circulated to the parish council and other interested authorities, and neighbours will be advised that plans are available for inspection.

Neighbours are always concerned about any effect on their own property, especially if any extension might affect their view. In theory the right to a view is not acknowledged as a planning consideration, because this could stop almost any development, but the term 'overbearing' is often used as an objection. It is possible to see and comment on letters of objection. It is helpful to have letters in support, so it is worth canvassing neighbours.

### In the event of problems

If there are objections to a scheme it is worth considering making amendments. If it seems that an application will be refused then it is possible to ask for a scheme to go to committee. An applicant cannot do this directly; a request has to be via a committee member. A list of committee members can be obtained from the council and the request for a hearing made to a local member. In some local planning authorities it is possible to speak to the committee for a few minutes. Informing the members about the scheme beforehand might help, simply because of the number of schemes they have to consider in a very short time.

### Outcome

If a scheme is approved the conditions should be checked carefully. If they are unsatisfactory or the application is refused, consider making an appeal to the DoE. See section on appeals and strategy before appealing.

## 5.3.6 Notes on planning approvals

### Duration

Planning permissions generally allow for work to start within five years of approval. There is no time limit for completion. Five years seems ample at first but delays can occur and planning policies sometimes change, so it is important to ensure that a start is made and that it is efficiently recorded, otherwise the permission runs out. It would probably be renewed but there might be a change of policy in the interim. In some cases the LPA reduces the time for commencement to less than five years. If so, it is necessary to act fast or appeal against the condition.

### Amendments

It is often necessary to make changes to an approved scheme either before or during building works. At this stage LPA approval is needed for any amendments. Once a house is finished some alterations can be made, for example to doors and windows, as for any existing house unless it is a listed building or subject to an Article 4 direction, private covenant or other restriction.

### Retrospective applications and acquiring planning permission

It is not illegal to undertake development without first applying for planning permission, except in the case of listed buildings. Planning permission can be applied for retrospectively. However, it would be unwise because there is always the risk that it will be refused and enforcement action taken. This might even result in a building being demolished and considerable costs.

Enforcement action by the LPA has to be instigated within four years of the offence in the case of buildings or 10 years in the event of change of use or failure to comply with conditions. Appeals can be made to the DoE against enforcement action. A guide to this procedure is available from the DoE in Bristol. A copy should be sent by the LPA in the event of an enforcement notice. Enforcement actions are complicated and it is advisable to seek professional assistance.

## 5.3.7 *Outline planning applications*

The purpose of an outline application is to establish the principle of development. It is often used to establish the value of a property prior to sale.

Outline applications are simple to prepare because detailed drawings are not needed, only site plans and the relevant certificate, and costs should therefore be low. However:

- The LPA may consider the site difficult and need drawings to prove that the new property would not harm existing ones.
- It can take as much time to consider an outline application as a full one.
- Outline applications are not acceptable for a listed building or land within its curtilage.

## 5.3.8 *Planning appeals*

Appeals can be made to the DoE against the refusal of planning permission, conditions which are unacceptable or non-determination, which is when the LPA does not reach a decision within eight weeks of the receipt of the application or any agreed extension. Appeals against a refusal have to be made within six months of the date of the refusal.

### *Appeal procedure*

A booklet by the DoE entitled *A Guide to Planning Appeals* clearly explains procedures and is available from the LPA or the DoE in Bristol, or the Welsh Office in Cardiff. Similar guidelines are available from the Scottish Office and the Northern Ireland Office. Appeals are conducted either by written representations, an informal hearing or a public inquiry.

Written representations are quicker and can often be conducted fairly easily by a private individual but the success rate is generally lower than for hearings or inquiries. If both parties agree, the appeal can be determined by the written representation procedure but either the appellant or the LPA can ask for an informal hearing or inquiry. Sometimes the DoE decide that a public inquiry is necessary.

Note that there are no fees payable to the DoE for planning appeals but there are for enforcement appeals. Professional fees may be incurred. Normally each party to an appeal pays their own costs but in exceptional circumstances costs may be demanded from an appellant or the LPA. This is explained in a pamphlet from the DoE, *Costs Awards in Planning Appeals*.

### Strategy before appealing

When an application has been refused it is advisable to attempt to get approval for an alternative scheme. This is in order to establish the principle of development beyond what might be allowed as permitted development and to reduce the number of matters at dispute. The difference in volume between the two schemes may be very little and the style a subjective matter of opinion over which it would be possible to appeal. In addition, a planning approval, whatever the style, will add to the value of the property. There is usually no fee for a second application within one year.

## 5.4　Building Regulations

### 5.4.1　What are they?

The Building Regulations set out the legal requirements for building work in England and Wales (Scotland and Northern Ireland have similar but different regulations). Introduced in 1965, they have always been concerned with the health and safety of people in and around buildings but, more recently, they have introduced more sophisticated measures for energy conservation. The Regulations are published in several parts, each with an 'Approved Document' which not only explains the legal requirements for that part but also gives practical or technical guidance on how to meet them. The Regulations were extensively revised and rewritten in 1985 and 1992. The Approved Documents are as follows:

| | |
|---|---|
| A | Structure |
| B | Fire safety |
| C | Site preparation and resistance to moisture |
| D | Toxic substances |
| E | Resistance to the passage of sound |
| F | Ventilation |
| G | Hygiene |
| H | Drainage and waste disposal |
| J | Heat producing appliances |
| K | Stairs, ramps and guards |
| L | Conservation of fuel and power |
| M | Access and facilities for disabled people |
| N | Glazing – materials and protection |
| Reg 7 | Materials and workmanship |

The Regulations do not apply to all buildings. Exemptions include greenhouses, temporary buildings, mobile homes, conservatories, sheds and detached garages with floor area less than 30 m². In the case of a garage it must be built substantially of non-combustible material or be at least 1 m from the boundary of the property to be exempt. A complete list may be found in Appendix 2 to the Regulations.

### 5.4.2 *Material alterations*

The Regulations do apply to 'material alterations' which are defined as follows.

If an alteration to a building, or any part of the work involved, would at any time 'adversely affect' the existing building as regards Part A, B1, B3 or B4 unless other work were done it is a 'material alteration' subject to control. The words 'at any time' refer to the work under construction; if the work contravenes during construction then a notice of intent must be given. Examples of such alterations are:

- removing part of a load-bearing wall which consequently requires the insertion of a beam to carry the load;
- altering a three-storey house in such a way that additional work is necessary to maintain the means of escape from the third storey;
- removing part of a wall which is non-load bearing but is necessary for fire resistance.

Inserting insulating material into an existing cavity wall or underpinning a structurally defective wall is also deemed to be material alteration. An alteration which affects disabled access is regarded as a material one for public buildings but not for dwellings.

### 5.4.3 Material change of use

The Regulations also apply when certain changes of use are intended (note – this is not the same requirement as that under planning law for 'change of use'). A 'material change of use' occurs in the following conversions:

- The building is used for the purposes of a dwelling, where previously it was not.
- The building contains a flat, where previously it did not.
- The building is used as a hotel or institution, where previously it was not.
- The building is a public building, where previously it was not.
- An exempt building becomes a building (e.g. a shed is converted to a bedroom).

A common example of a material change of use would be the conversion of a Victorian terraced house into flats as illustrated in Figure 5.1. The height of such a building is important because it affects the provisions for means of escape (Part B). The uppermost floor is less than 11 m high, which allows the use of a single stair escape route, but the number of storeys requires a fairly stringent requirement for the fire resistance of the structure, in this case almost certainly a period of one hour.

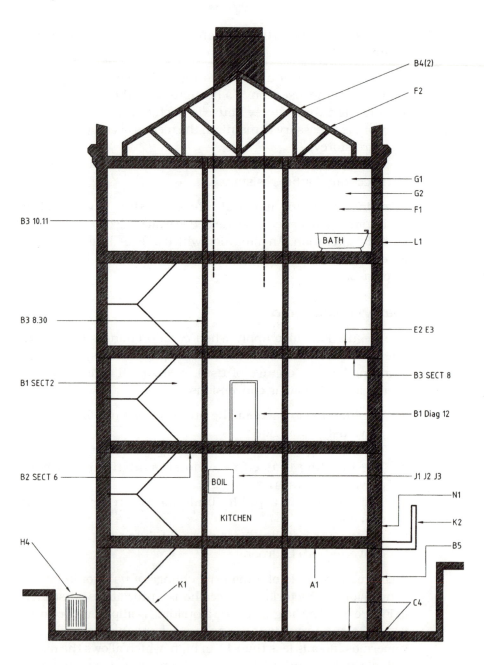

*Figure 5.1    Building regulations applicable to a Victorian house conversion into flats*

Figure 5.1 gives some indication of where the most important regulations would apply. The following notes are only briefly explanatory. The reader is recommended to read the Approved Documents for more detail:

**Part**

**A** Additional loads may require an investigation of foundations – unlikely to be needed in this case; more applicable to hotels, institutions and public buildings.

**B1** (Means of escape) – people escaping from the second floor and above must have a safe, protected route to the ground level and out. The existing staircase would normally need to be enclosed.

**B2** (Internal firespread – surfaces) – surfaces within a protected escape route must be incombustible so that fire cannot spread across them. Wood panelling to be avoided.

**B3** (Internal firespread – structure) – the walls and floors surrounding an escape route and between flats must have a degree of fire resistance as stated above. Plaster and plasterboard have good fire resistance.

**B4** (External firespread) – buildings close to one another must be protected from firespread by having fire resistant cladding. Terraced houses are particularly at risk but tiles and slates can cope with this.

**B5** (Access for Fire Service) – the basement should have ventilation to remove heat and smoke in the event of fire.

**C** (Resistance to moisture) – walls should have damp proof courses. If built after about 1880 this building may already have them.

**E** (Sound insulation) – walls between dwellings should resist the transmission of airborne sound. Floors should resist the transmission of airborne and impact sound. Victorian builders of any quality tended to use thick dense materials which are good sound insulators. Lath and plaster ceilings are typical and should be retained wherever possible.

**F1/F2** (Ventilation) – roof spaces and bathrooms must be ventilated, in the latter case by mechanical means. The

155

fundamental purpose here is to remove stale and damp air which can cause materials to deteriorate and condensation to occur.

**G1** (Food storage) – storage of food should be in 'adequate accommodation'. Refrigerators are deemed adequate but the value of preserving an existing larder should not be overlooked.

**G2** (Bathrooms) – dwellings must have bathrooms (with a bath or shower) with a supply of hot and cold water.

**G4** (Sanitary conveniences) – all buildings should have them and they should be 'designed and installed so as to allow effective cleaning'.

**H4** (Solid waste storage) – refuse containers are necessary and must be accessible from a street.

**J1–J3** (Heat producing appliances) – boilers and heaters must be properly installed, particularly with respect to air inlet and exhaust.

### 5.4.4 Extensions

Extensions can be upwards, sideways or downwards. In most respects they are treated as new buildings. Not all of them need to comply with the Building Regulations however. Exemptions include:

- greenhouses, conservatories or porches with a floor area of less than 30 m²;
- car ports, open on at least two sides with a floor area of less than 30 m². Most single car ports are about 18 m² in plan;
- covered ways or yards with a covered area of less than 30 m².

Some small buildings such as garages are not subject to control if they are detached but if they are built within 1 m of a boundary they must be built substantially of non-combustible material. If they are extensions they must comply in full. Conservatories and glazed porches are only exempt if they have safe or protected glazing below a height of 800 mm generally but 1500 mm in doors and side panels to doors.

As a general rule extensions should be carried out so that the original building is not adversely affected in relation to any of the Regulations. Certain elements in the original building may not comply – what the extension must not do is make matters worse.

With regard to thermal insulation (Part L), if the extension is less than 10 m² in plan then it can be constructed in a similar way to the existing construction. If larger it must be insulated unless the option of calculating energy consumption levels is being taken.

### 5.4.5  Loft conversions

Apart from a general requirement that they must comply, loft conversions are specifically referred to in Parts B (fire safety), K (stairs, ramps and guards) and L (conservation of fuel and power). Briefly the requirements are as follows.

***Part B***

In an existing two-storey house and provided the roof line of the conversion is not raised above the original, if the floor area of the loft is less than 50 m² and the loft is not more than two habitable rooms, then the work is a loft conversion rather than a three-storey house. In this case:

- The staircase must be enclosed with fire resisting walls and should extend to a final exit or two alternative routes to a final exit.
- New doors to stair to be self-closing half-hour fire doors. Existing doors may be just self-closing.
- Glazing in stair enclosure to be fire resisting.
- The new stair to comply with Part K (see below).
- Loft floor to have a minimum of half-hour fire resistance.
- Escape may be through a window in the roof, with a clear opening of 850 × 500 mm.

***Part K***

The staircase to the loft should comply with the following:

- Where 2 m headroom (the normal requirement) is not possible, 1.9 m at the centre tapering to 1.8 m at the edge will be acceptable.

- The rules for spiral stairs may be less stringent than normal.
- 'Alternating tread' stairs are acceptable where space is insufficient for standard stairs.
- A fixed ladder may be used if it provides access to only one habitable room.
- Retractable ladders are not permitted for means of escape.

**Part L**

In the case of loft conversions, the thermal insulation requirements are less onerous than for a normal domestic roof. In the latter case a maximum 'U' value of 0.25 W/m² K is required whereas for the loft conversion it can be increased to 0.35 W/m² K.

### 5.4.6   Getting approval

The Building Regulations are administered by the local authority or an approved inspector. If a local authority is used then an application for approval can be made in one of two ways:

1   *Deposit full plans* – full detailed drawings are submitted in advance of any work. These drawings are then passed or rejected by the local authority. In the latter case it is usually a question of providing further information or making amendments to satisfy the local authority. This is very common practice and should not be a cause for alarm.

    Drawings are submitted in duplicate, together with a site plan if new ground is to be covered, at a scale sufficient to allow the site to be easily identified with the site coloured in red. A copy of a map of the area may be photocopied in the local reference library or traced from an Ordnance Survey map available on microfiche from some booksellers. To photocopy an Ordnance Survey map is an infringement of copyright.

    Maps and plans produced by others, as sales particulars for example, should be used with caution and may require a fee to be paid to the originators or at least their permission sought.

2   *Give a Building Notice* – no plans are required unless a new building is proposed, and then only a block plan is needed. This method is suitable for all buildings except offices, shops and hotels. There is a risk in this case that the work, not having been checked in advance by the local authority, will not comply and have to be redone. Building Notices do not

apply to buildings such as shops and offices where the Fire Precautions Act 1971 applies.

The advantage of the full plans route is that one benefits from getting the prior advice of the building control officer. In both the above methods it is necessary to give the local building control office two days' notice of starting work. It may be necessary to provide the local authority with structural calculations where alterations to main structural elements are to be carried out. Normally these would be done by a civil or structural engineer, although where prefabricated components such as lintels or roof trusses are to be used, calculations should be obtainable from the manufacturer.

In the event that a particular regulation is too onerous or cannot reasonably be satisfied, the local authority may agree to dispense with it. This is rarely necessary but there have been cases when dealing with listed buildings, for example, where a heavy timber frame does not comply because it is combustible but in practice is accepted as having a satisfactory level of fire resistance. In the event of a disagreement between the local authority and the applicant, the matter can be referred to the Secretary of State for determination.

There has been some confusion in the past about the need to submit separate applications for Buildings Regulation and Planning approvals. Where a proposal is permitted development as defined in Planning (see Part 5.3.2) an owner need only apply for Building Regulation approval, but because the application has been notified to the planning department (now common practice) he or she will receive a letter claiming that planning permission is required. There is no point in paying fees or wasting the planning authority's time unless permission is actually required, so to avoid the situation a letter explaining in terms of size, location etc. how the proposal constitutes permitted development should be included with the Building Regulations application.

## 5.4.7 Approved inspectors

There is an alternative way of obtaining approval under the Regulations. This is to enlist the services of an 'approved inspector', usually an experienced consultant or an inspector certified by the government. Such a person will check drawings for compliance, issue a plans certificate confirming approval, inspect the work as it proceeds and finally issue a certificate of completion. Normally, a fee is payable to the local authority for approval under the Regulations but if an approved inspector is employed, the fee goes to that person. The fee is based on the number of dwellings and would amount to about £305 including VAT for a new house built in 1998.

If building work is carried out which does not comply with the Regulations, the inspector or local authority will ask for it to be altered or removed. If the owner refuses to do this the local authority may serve a notice requiring the remedial work to be done. If the owner continues to object, further action will take place in the Magistrates' Court where both sides will need to convince the magistrates of their case.

A list of approved inspectors current at October 1997 is included in Appendix C.

## 5.5   CDM Regulations

### 5.5.1   What are they?

The Construction (Design and Management) Regulations 1994 (CDM Regulations) came into force on 31 March 1995. They are primarily concerned with construction safety but unlike previous legislation they place the responsibility for safety on clients and designers as well as contractors and sub-contractors. The principal requirements under the Regulations can be summarized as follows:

*The client is to*:

- Appoint a 'planning supervisor' who must co-ordinate the health and safety aspects of design, prepare a pre-tender stage health and safety plan and ensure that a 'health and safety file' is prepared. These are discussed below.
- Provide information on health and safety to the planning supervisor. This is an important aspect of briefing. Information about a building, a site, a use or activity which may affect health or safety must be included.

- Appoint a 'principal contractor' before construction starts so that they can carry out or manage the health and safety requirements on site.
- Ensure that those who are appointed to these posts are competent and suitably equipped to carry out their duties.
- Ensure that the health and safety plan is prepared before construction work starts.
- Ensure that the health and safety file is made available at the end of the job. This is a document for use by those who maintain, repair or alter the building later. It provides information, for example, on how high-level windows should be cleaned or how frequently a boiler needs to be serviced.

*The designer is to:*

- Advise clients who may be unaware of their duties under the Regulations what must be done as outlined above.
- Design in such a way as to avoid or control  the risks to safety, health and welfare of people during construction and later in the maintenance, repair and use of the building.
- Provide full information about the health and safety risks in the design to others particularly the planning supervisor, the principal contractor and anyone involved with constructing the design.
- Co-operate with the planning supervisor and other designers, particularly where there is an overlap in design responsibility such as would occur, say, in the design of a complex services installation.

**The health and safety plan**

- Is about the safety management of the project.
- Should be in proportion to the nature, size and level of risks involved in the project. Small jobs require only small plans.
- Should include information from the client and designers.
- Should be prepared before work starts but be updated as work proceeds.
- Should highlight significant risks which cannot be avoided.
- Should point out health hazards arising from the use of materials.

- Should consider the hazards in the use of vehicles, plant and equipment on and off the site.
- May be a question of giving emphasis to information which is otherwise included in tender documents, such as specifications or bills of quantities.
- Will initially be the responsibility of the planning supervisor but the principal contractor when appointed will have a significant input, as will any sub-contractors involved.

### The health and safety file

- Is rather like a car owner's manual for buildings.
- As with the plan, should be in proportion to the scale and complexity of the structure. Simple buildings require only simple files.
- Should grow as the job proceeds, gathering all information pertaining to the safe use of the completed building.
- Will include information from the principal contractor on services, plant and equipment installed in the building.
- Will be organized and co-ordinated by the planning supervisor who must pass the file to the client as soon as reasonably possible after completion.
- Must be kept by the client for use by all who use, repair, alter or maintain the building.
- Can be in two parts, one dealing with day-to-day maintenance, the other including information such as drawings for use in later alteration work or major repairs.

## 5.5.2 The application of CDM

Contractors have carried the responsibility for safety, health and welfare in building for many years. The law has decreed it so, in particular:

The Construction Regulations of 1961 and 1966 (now largely replaced)
The Health and Safety at Work etc. Act of 1974 (HASWA 74)
The Management of Health and Safety at Work Regulations of 1992
The Construction (Health, Safety and Welfare) Regulations 1996

The CDM Regulations represent an important change. For the first time, clients and designers have expressly stated duties and responsibilities, and the safe use of the building after completion is included. While it is true that under Section 6 of HASWA 74 designers of machinery and equipment had to design with safety in mind, this seems never to have applied to the designers of buildings.

The Regulations require projects lasting more than 30 days or involving more than 500 person days of work to be notified to the Health and Safety Executive and all the regulations will apply. If the project is not notifiable, the Regulations will still apply if more than 4 persons are working together or if any kind of demolition is involved. Works for a 'domestic' client are not affected by most regulations unless they can be classed as speculative development. Of significant importance is the fact that designers must conform totally, regardless of the size or nature of the contract.

Fundamental to CDM is the practice of risk assessment and management. The Regulations do not say that risks can be eliminated, only that they should be recognized and assessed. Designers (in the broad sense that they are defined in the Regulations) have a particular and relatively new responsibility to balance the costs and benefits of safe design. To design, for example, the facade of a building in such a way that it is virtually maintenance free, and therefore safer, may require sacrifices to be made in terms of cost and appearance. Earlier (Part 3.3.5) we saw that design has a profound effect on maintenance costs. To that must now be added its effect on safety in maintenance.

Anyone can be appointed a planning supervisor or principal contractor but it makes sense if they are the design team leader (architect, surveyor or engineer) and the general contractor. Either or both could in fact be the client. What is required, though, is that they should be competent and have the resources to do the job. As discussed in Chapter 4 members of professional institutions can be expected to be broadly competent but whether they have yet become familiar with CDM is open to question. Similarly, contractors may well be experienced but have not fully appreciated what CDM requires of them. The client should be aware perhaps of such possibilities.

Under the CDM Regulations, design includes preparing drawings, design details, specifications and bills of quantities. It would seem that very little design work can escape, particularly since 'specification' in its broadest sense covers a wide range of activities.

The Health and Safety Executive (HSE) is responsible for enforcing the Regulations and is the body to which the planning supervisor should give notice that a building job is about to start. Any questions concerning the Regulations should be addressed to the local office of the HSE which can also supply relevant literature. Failing that, contact the HSE Information Centre at Broad Lane, Sheffield S3 7HQ, tel: 0114 289 2345.

A very positive aspect of the Regulations is the strong emphasis they place on an exchange of information and therefore good communication between designers, between contractors, between designers and contractors and between clients and their agents.

Another is that although many small jobs will be exempt, nevertheless the establishment of new procedures, new ways of working and the concern for health and safety in detail will surely have a positive effect on all construction work in the future.

# Further reading

Bryan, H. (1996) *Planning Applications and Appeals*. Butterworth.

Department of the Environment (DoE) (1997) *Planning: A guide for householders*. HMSO.

Department of the Environment (DoE) (1997) *A Guide to Planning Appeals*. HMSO.

Department of the Environment (DoE) (1997) *The Party Wall etc. Act 1996 explanatory booklet*. HMSO.

The Lord Chancellor's Department (1995) *Resolving Disputes Without Going to Court*. The Lord Chancellor's Department.

Polley, S. (1995) *Understanding the Building Regulations*. Spon.

Uff, J. (1996) *Construction Law* (6th Ed). Sweet and Maxwell.

# 6

# Updating houses

This chapter describes some of the characteristics of Victorian terraced houses and the ways in which they have been 'improved' in the past. It shows how changing trends can stimulate the need to improve a home, particularly with respect to the kitchen, the bathroom, thermal insulation and heating.

Updating and repair are continual and essential processes. A report by the RICS in September 1977 pointed out that one quarter of all housing stock in the UK was built before the First World War and that not enough repair work is undertaken. A large proportion of the pre-1914 stock was built in the Victorian era and so becomes a focus for this book.

## 6.1 Victorian terraces

Much residential property close to town centres consists of terraced houses. They were built for a different lifestyle and to different standards of construction, heating and insulation than are acceptable today.

Some of these houses have been well maintained and sympathetically adapted, some have not. So there are three cases for updating:

1  *An unchanged house that has been well maintained.* These are rare and may command a premium. Updating may involve conflict with ideals of conservation.
2  *A poorly 'modernized' house.* The 'new' work is often already outdated and the original character has been lost. This tends to depress the market price and some restoration of character, as well as updating, will help to improve marketability.

3  *An unchanged house in poor repair* allows an opportunity to improve standards and retain most of the character. In a normal housing market the asking price should allow scope for refitting:

Electricity, gas and water services

and considering:

Windows
Fireplaces
Staircases
Relationship to garden

and updating:

Kitchens
Bathrooms
Space and water heating
Thermal insulation

## 6.2  The original Victorian house

A typical well-maintained unchanged terraced house would have the following accommodation:

● A WC approached from a backyard, no bathroom
● A small kitchen with a larder cupboard and stove or bread oven with a chimney
● A family-cum-dining room next to the kitchen at the rear
● A parlour or best room at the front – used mainly on Sunday
● Three bedrooms directly above the ground floor rooms
● Chimney breasts with fireplaces in each room
● Trap door access to the loft, or narrow staircase to an attic floor in larger houses
● Cellars under the main reception rooms in some cases

The width of the houses would have varied from 3 m to 5.5 m. See Figure 6.1 for a house of this type.

Features and finishes usually include:

● Timber boarded floors except for kitchen and cellar which would be brick, tile or stone flags

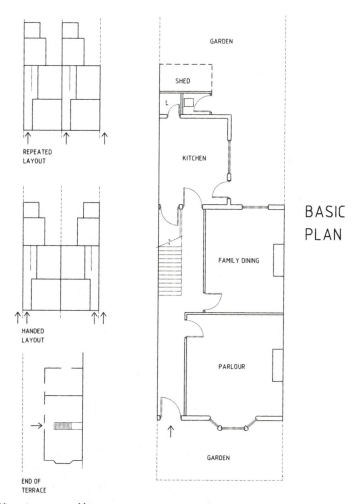

*Figure 6.1   A typical Victorian terraced house*

- Panelled doors
- Balusters
- Fireplaces in cast iron, with wood or stone mantelshelf

The amount of elaboration varies with the size and quality of the house. The smallest and simplest would have plain doors and balusters without any mouldings. Larger houses usually have turned balusters, mouldings on both sides of door panels, tiled surrounds to fireplaces, decorative plaster cornices, moulded picture rails and skirtings. Door locks are generally

167

Figure 6.2   *Some Victorian details (after Hugh Lauder and Alan Johnson – see Further reading)*

surface mounted with brass knobs and finger plates. See Figure 6.2. Daylight to the hall and landing is usually provided by a glazed panel over the front door and the rear bedroom. Windows are timber vertical sliding sashes, counter balanced by cast iron weights in sash boxes, sometimes built into a canted or splay sided bay with panelled shutters.

### *Early modernization*

The first change to many of these houses was to make the WC accessible from inside the house. The external doorway was bricked in and a new doorway made into the kitchen.

The second change was usually to dispense with the larder and turn the whole area into a small bathroom. Then building regulations changed and it was no longer possible to have a WC opening directly into a kitchen. It is interesting to note that the regulations have now changed back again with the use of mechanical ventilation.

## 6.3   The poorly 'modernized' house

The alterations most unsympathetic to older houses were done in the 1950s and 1960s. It was an era with a pent-up urge for change and a passion for streamlined design. Shortage of materials after 1945 stifled innovation until the 1950s when the post-war Festival of Britain in 1951, on the 100th anniversary of the Great Exhibition at Crystal Palace, stimulated a wave of interest in modern design. Surfaces had to be smooth and flush, doors and staircase were panelled with hardboard, chipped earthenware belfast sinks and stained teak draining boards were changed for stainless steel sinks and 'formica' worktops. Easy to apply emulsion paint replaced traditional oil paint, limewash and distemper.

### *6.3.1   Fireplaces*

Fireplaces suffered the worst internally for two reasons.

1   The proportions of fireplaces looked all wrong to the modern eye. They were too tall and sedate, quite out of keeping with the low streamlined image that was wanted.
2   Fireplaces no longer had a function. Smokeless fuel became mandatory in most urban areas and it produced a less attractive fire. Central heating had taken over as the ideal form of space heating.

Fireplaces were often:

● Changed for up-to-date versions (those of the 1930s now look stridently out of place)

- Removed and the opening sealed
- Removed together with the chimney breast, sometimes to the impairment of the structure. The space occupied by the chimney breast was often required for fitted wall-to-wall wardrobes in the bedroom.

Fireplaces on the ground floor are the ones usually altered. If they are still intact on the first floor it is worth moving a feature fireplace downstairs, where it will be more widely seen. Fireplaces on the first floor are generally not worth restoring because they are not used and disrupt the organization of a bedroom, unless the large front bedroom becomes a living room.

Fireplaces almost seem to have a symbolic status, and definitely sell houses.

### 6.3.2 Windows

Windows suffered the most externally from 'modernization'.

Steel windows were the first to replace timber, but there are relatively few in terraced houses because they were manufactured in standard sizes with a horizontal emphasis which was not adaptable to the old window openings. Where they were installed by altering the openings, many rusted when the galvanizing failed.

Aluminium was the new wonder material for windows because it could be 'anodized' and did not need painting. There was a particular vogue for louvred windows, which gave a modern horizontal emphasis even to vertical openings. Horizontal and vertical aluminium sliding sashes with slender structural members were installed instead of timber. This fashion passed because the windows had poor insulation qualities and the aluminium corroded in polluted urban atmospheres.

Both steel and aluminium windows are a poor substitute for timber, in terms of both appearance and performance. The frames, sashes and glazing bars are too flimsy to look right in a Victorian building, and their insulation qualities are low.

PVCu windows became the next fashion in fenestration. These have the advantage that they can be manufactured to match existing openings, and are double glazed. The better ones match the original glazing bars and fold down in order to be safe and easy to clean from inside. They do not need painting, which is the main selling point. However, they tend to look artificial, particularly where they incorporate small drainage channels at the bottom of the sashes, and the thickness of the glazing bars rarely looks right. The worst design in PVCu windows is that which has a large fixed lower pane and a small vent above. When these windows are double glazed it is almost impossible to break them open to escape from fire. They are greatly disliked by conservationists.

### 6.3.3 Restoration after poor modernization

The most usual jobs are to:

1 Restore the front door to original or appropriate style
2 Replace original fireplaces on the ground floor
3 Remove hardboard from doors and staircase
4 Replace vinyl asbestos tiles, preferably with terracotta or other ceramic tiles in rear kitchen area
5 Change sanitary fittings for style to suit house
6 Replace timber sash windows

Small joinery works and some of the large DIY chains will make vertical timber sliding sash windows to fit existing openings. These incorporate double glazing and security locks, and most can be opened inwards for easy cleaning.

## 6.4 Changing trends

- Smaller households
- Extensively equipped kitchens
- Larger reception rooms, linked with the garden
- Showers in place of baths

### 6.4.1 Smaller households

General trends to smaller families and one or two people households. More demand for one or two bedroom dwellings, particularly for letting purposes, because the initial cost and wear and tear are lower for the landlord, and the rent is less for the tenant, reducing the need to share. The outcome is the loss of a third bedroom to a bathroom, particularly if a laundry area can be installed, but it may not reduce value as before. The standard of finish tends to be more important.

### 6.4.2 Extensively equipped kitchens

More space is now needed for a wide range of labour saving equipment:

- Dishwashers, clothes washing and drying machines
- Microwave and split-level cookers
- Refrigerators and freezers
- Food mixers and liquidizers
- Ironing tables and flat bed irons
- Vacuum cleaning and carpet shampooing
- Gadgets such as toasters, grills and griddles

See Section 6.5 for a further discussion on kitchens.

### 6.4.3 Larger reception rooms

The trend for larger reception rooms is often met by forming an opening between front and rear rooms. This can create a pleasant spacious open-plan effect with views through to the garden. If the kitchen is small this also becomes a dining kitchen and even a children's play room. However, it reduces the flexibility of use – it is less easy, for example, to use one room as a study or work room, or even for letting. Forming large openings will also affect the original character. Folding or sliding doors in the centre is a way of having the advantage of both layouts, but soundproofing has to be considered. There are firms which specialize in producing soundproof room dividers.

For other ways of creating a larger living room, see Chapter 7.

### 6.4.4 Showers in place of baths

Spacious houses have room for both but there is an increasing tendency for tenants to request shower facilities. En suite facilities too are in demand.

These changes save time, water and space, but showers are more costly than baths to install and great care has to be taken over leakages. The changes may have repercussions on the plan, as we shall see in Chapter 7.

# 6.5 Kitchens

## 6.5.1 *Location*

Kitchens do not have to be relegated to the back of the house where they are usually separated by one or two steps from the dining room. They can be in the front so that the cook can overlook the street or in the centre for family dining. The rear room still makes a good position for a galley-type kitchen but there is not usually enough space for a dining table.

### *Alternatives*

1 At the front. See Figure 6.3. The main advantage is that it is easy to keep in touch with events outside, and so is particularly useful for people confined to the house. One disadvantage is that the front room and its fireplace are the most important for establishing the character of the house and it may be unwise to disrupt this.

2 In the centre. See Figure 6.4. A very comfortable location. A disadvantage of this is that a reception room is partially lost – unless the cook is very tidy. This can be compensated for by extending the kitchen as a living room, a possibility discussed in Chapter 7.

## 6.5.2 *Kitchen fittings*

Refitting a kitchen is complex; it requires co-ordination of services, plumbing, gas and electricity for cooking, refrigeration and waste disposal and cabinet fitting for sinks and storage units etc. DIY chains which supply the storage units and equipment will also help with planning the layout and may fit the whole installation as required. Specialist kitchen fitters will design and install kitchens to individual requirements. Some also manufacture storage units and so offer flexibility in worktop heights and unit widths.

### *Size and layout*

These can vary enormously. Mini-kitchens which incorporate a hob, sink, microwave and refrigeration in units from 1 m to

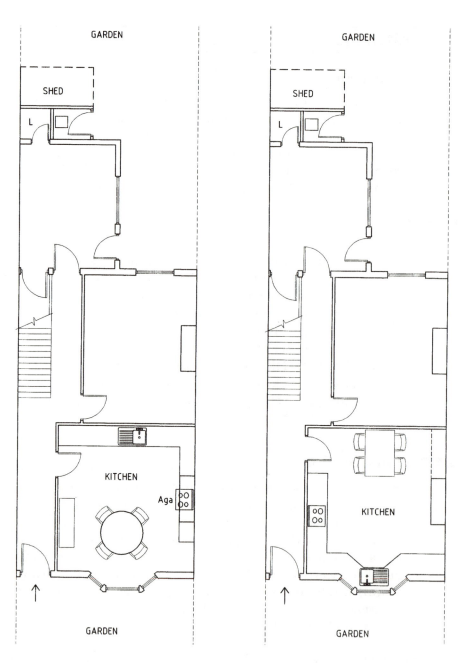

*Figure 6.3   Alternative plans for front kitchens*

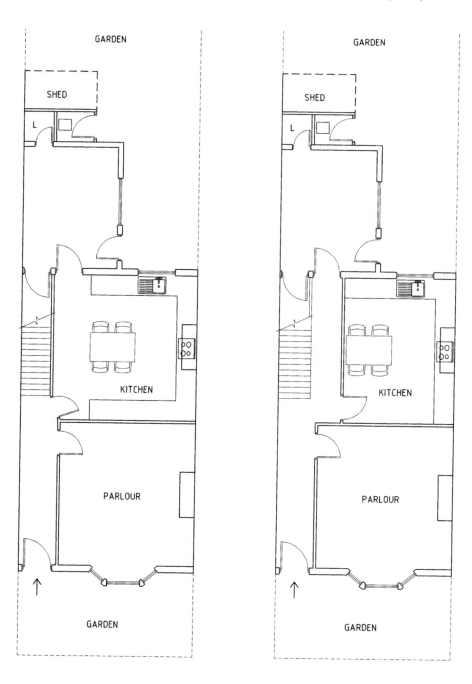

*Figure 6.4   Alternative plans for centre kitchens*

1.5 m wide are suitable for a single person studio flat. See Figure 6.5. At the other extreme is a family dining kitchen with room for a deep freeze, double oven, dishwasher and ample storage for food and crockery.

The main points to note are:

- The most used worktop is the one between the sink and the hob.
- Elbow room is needed on both sides of a hob.
- Lighting is important for safety and efficiency.
- A galley-style layout is efficient but room for a breakfast table is desirable.
- Island units need careful pre-planning to run services under the floor.

### Choice of units

Most units are supplied as flat packs and are relatively easy to assemble.

Carcasses – i.e. framework and shelves – are generally the same throughout a manufacturer's range and the standard has improved recently. The cost differences come in the style and finish of the cupboard doors. It is advisable to wait for retailers' special offers to rotate. White is probably the most consistently popular colour for units but the durability of the finish is important. Constant use, grease, spills and knocks all affect the surfaces so they need to be hard-wearing.

Spills or leaks in sink cupboards or excess water on the floor will soak into the composite boarding of the carcass and cause it to swell. Silicon sealant along the internal joints and at the junction of the plinth with the floor will help to protect the units.

KOF1500. As KO1500 But also includes 1 x 1500mm backboard and 1500mm run of wall units.
£1462.00 excluding VAT.

*Figure 6.5   A Neff mini-kitchen*

### Cookers

Eye-level ovens and separate hobs avoid the hazard of lifting heavy casseroles but the cooking performance in the same manufacturer's range, such as New World gas cookers, hardly varies; so the key considerations are:

- Appearance
- Available space
- Cost

In terms of cost and space the most economic are free-standing cookers with an eye-level grill and storage space below the oven. Separate microwave ovens are also the most cost-effective and simple to use, but need more space than combination ones.

### Worktops

- Granite is superb in terms of appearance and wear, and is stain resistant, but very expensive.
- Wood block looks attractive but requires regular maintenance.
- Melamine laminated surfaces also wear well and are readily available at all DIY merchants. White looks crisp and is easy to keep clean with one of the many kitchen cleaners available.

### Floors

- Need to be hard-wearing and slip resistant.
- Water and grease on a kitchen floor can make it quite dangerous, so the surface needs to be matt or slightly grainy.
- There is a wide choice of ceramic tiles generally available.
- Terracotta tiles are ideal, preferably on a solid floor base but any dropped crockery will break.
- Cork-o-plast is more resilient and cork has a timeless quality. As with heavy duty vinyl tiles, the surfaces look shiny but have a slight texture.
- Marble is not suitable, it is slippery and stains rather easily.
- Cleaning and maintenance of a kitchen floor are important considerations; it is always advisable to seek the advice of the manufacturer or supplier on the best and cheapest methods to use.

### References

MFI, B&Q, Wickes and other retailers produce leaflets or brochures on their kitchens. Some are design sales oriented, others more basic, but they are all backed with in-store assistance. The leaflet produced by Wickes, *Planning a Wickes Kitchen*, is particularly helpful in discussing the layout and drawing attention to safety factors. Graph paper and standard units drawn to scale facilitate planning. Mini-kitchens, manufactured by Neff, are available from Southway Interiors of London NW2, tel: 0181 452 8011.

## 6.6 Bathrooms

### 6.6.1 Location

Bathrooms and shower rooms are best located on the same floor as bedrooms to avoid very young or elderly people having to negotiate stairs at night. Usually this implies a first floor bathroom, but one of the ground floor rooms might be a bedroom, so a ground floor cloakroom is desirable.

### 6.6.2 Cloakroom possibilities

1 Understairs. See Figure 6.6(a). The reduced headroom under the stairs is usually not an obstacle because headroom is not needed less than 650 mm from the rear wall. The WC suite should not be close coupled, to allow an overflow to be connected to the flush pipe. Close coupled suites need overflows elsewhere.

2 If the house has a cellar this arrangement is not possible. The cloakroom then could be at the back of the kitchen, the position usually found in early modernizations. See Figure 6.6(b). It is acceptable again with mechanical ventilation but views from the kitchen are restricted to the yard and the house opposite.

3 Another possibility is to leave most of the plumbing intact, change the bath for a shower and give access to the garden. See Figure 6.6(c).

4 Relocating the cloakroom opposite the back door improves the relationship to the garden and facilitates adding an extension or conservatory. See Figure 6.6(d).

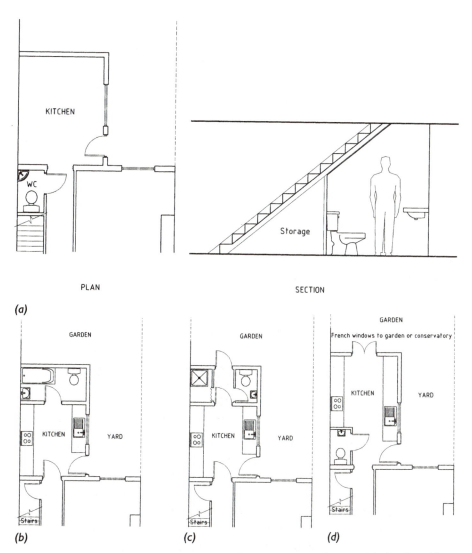

PLAN

SECTION

*(a)*

GARDEN

*(b)*

GARDEN

*(c)*

GARDEN

*(d)*

*Figure 6.6 (a) The under-stairs cloakroom; (b) The cloakroom to the rear of the kitchen; (c) Replacing the bath with a shower; (d) The cloakroom opposite the back door*

## 6.6.3 First floor bathrooms

### Rear bathroom

The standard solution has been to convert the rear bedroom to a bathroom. See Figure 6.7(a). This has advantages and disadvantages.

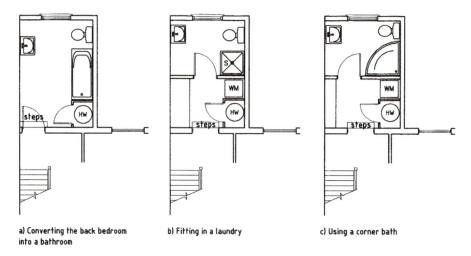

a) Converting the back bedroom into a bathroom

b) Fitting in a laundry

c) Using a corner bath

*Figure 6.7    Bathrooms*

|                 |                                                                 |
|-----------------|-----------------------------------------------------------------|
| Advantages:     | Large bathroom                                                  |
|                 | Localizes plumbing over kitchen                                 |
| Disadvantages:  | Bathroom larger than necessary                                  |
|                 | Loses a pleasant bedroom                                        |
|                 | Rarely on same level as other bed-rooms, so not particularly convenient |

As the bathroom is overlarge and the kitchen is not, one possibility is to change the bath for a shower and fit a laundry area in front of the bathroom, with space for the washing machine. A dryer above could be vented through the external wall. Space for ironing or cupboards would be created. See Figure 6.7(b). Corner baths tend to look comfortable and lavish but often can be fitted into small bathrooms and provide a more luxurious solution than a shower tray, but still with shower fittings. See Figure 6.7(c).

## Centre bathroom

See Figure 6.8. Installing a bathroom in the centre of the house leaves the bedroom with the most attractive relationship to the garden intact. Other advantages are:

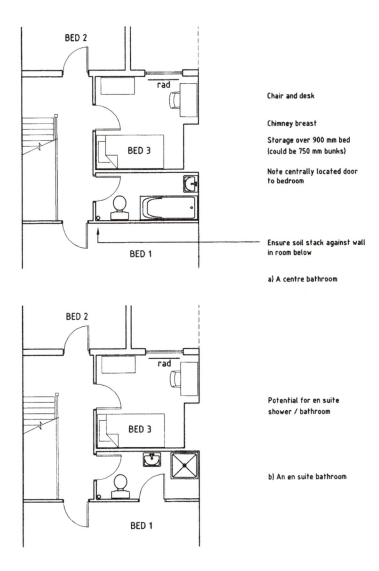

*Figure 6.8  Centre and en suite bathrooms*

- No outside walls – warmer location, less condensation.
- Still a third bedroom for a child, guest, or study room.
- Bathroom on same level as main bedroom.
- Possibility of en suite bathroom as in Figure 6.8 (b).
- However: bathroom and bedrooms are both small.

**Note**

The location of the doors is particularly important in a small bedroom, in order to allow space for a bed, storage, a chair and some furniture as shown in Figure 6.8. Space in small areas should be considered three-dimensionally, as in boat and caravan design. Bunk beds, raised beds with storage below or high-level cupboards all increase storage capacity.

### 6.6.4 Attic bathrooms

People step in and out of the centre of baths and only need full standing room close to the taps. So a bathroom can be set quite close to the eaves. A corner bath utilizes the space and limited headroom efficiently. Additional headroom can be provided by careful positioning of roof lights. See Figure 6.9.

### 6.6.5 Bathroom fittings

#### Choice and cost of sanitary ware

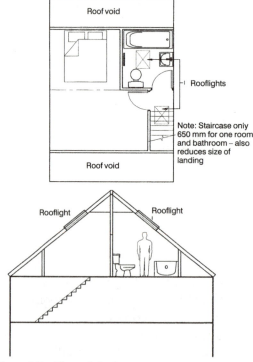

Figure 6.9    The attic bathroom

A very wide range is available from specialist shops, builders' merchants and DIY chains. The simplest and cheapest way is to buy a suite which includes a bath, side panel, washbasin, taps, waste fittings, plugs and chains, WC, handle and seat complete with a cover. Changing any of these items adds disproportionately to the cost. For example, in a suite costing £250 the WC bought as a single item could be £120. So it pays to buy the suite as a whole and add on any changes such as the type of taps. Surplus taps could always be used elsewhere or go to car boot sales, or baths be used for garden ponds. DIY chains often rotate their suites as special offers.

### Design

Choose sanitary ware which suits the style of the house, preferably in white. Colours go in and out of fashion and tend to date the era when the house was last modernized, so they may make a house look old-fashioned unless assembled with design flair as part of a comprehensive decor.

### Baths

Baths may be acrylic, enamelled steel or cast iron.

Acrylic baths retain the temperature of the water longest. Most are well designed to fit body contours, so they are comfortable to use and easy to clean when warm. However, they flex in use slightly, which may make the joint with the wall difficult to seal.

They are also vulnerable to cigarette burns, a factor to consider in rented property. For durability, acrylic baths should not be of material less than 5 mm thick. There are ranges of heavy duty acrylic baths but they are not usually found in suites.

Enamelled steel baths are a good rugged solution and are now manufactured complete with handgrips in more comfortable shapes.

Older baths are cast iron and many have classic shapes with elegant feet, now much prized. Many have been discarded because of badly pitted or worn surfaces. It is possible for them to be resurfaced in situ. See *Yellow Pages* for specialists. New cast iron baths are relatively expensive. The surface is not as good as in the past because lead is not used in the manufacturing process.

All baths should have overflows fitted to the waste.

### Bath panels

Acrylic panels come as standard with most suites. They deflect and crack rather easily and do not provide adequate toe space in a tightly planned bathroom. Painted or laminated plywood panels can be made quite easily with a good toe space incorporated.

### Washbasins

The ideal shape has ample flat surface at the side of the taps to act as a shelf close to the wall but fairly steep sloping sides so that the water drains away. In hard water areas, pools of water standing in the basin will lead to salt deposits which are difficult to clean. Pedestals have the advantage of concealing pipework as well as providing support.

### WCs

Close coupled suites are the most sculptural and stylized. Concealed cisterns have a clinical simplicity of appearance but are more difficult to install except for the DIY enthusiast. Cloakroom suites often have separate pans and cisterns connected by the flush pipe. Narrow plastic cisterns enable WCs to be fitted in tight spaces but again beware of cigarettes.

Overflows: WC cisterns have to have overflows, usually to an outside wall, but linking into the down flush pipe is acceptable. This is only possible where the pan and cistern are separate, not with close coupled suites.

Siphonic suites are no longer made by some major manufacturers. They were designed to have a more efficient clearing action but in some circumstances were more liable to blockage and have been replaced generally by standard wash down models. However, there is a wider range and sophistication of design, including models that are much quieter to operate and use less water, now on the market.

## 6.7 Thermal insulation

Areas to consider are the roof, the ground floor, external walls, external doors, disused flues and windows.

Ventilation has to be considered at the same time as insulation. Some of these areas should be insulated as soon as possible especially the roof which may account for 25 per cent of the total heat loss. Others can be included in other repair work, such as external walls (35 per cent) and ground floors (15 per cent). Windows and ventilation usually account for 25 per cent of the total loss.

### Roofs

Insulate between ceiling joists if the roof space is not used as an attic. Otherwise insulate between the rafters. In the latter case 50 mm ventilation space has to be allowed above the insulation for air to circulate over the rafters so materials which provide a high degree of insulation without great bulk, such as foil-encased extruded polystyrene boards, are most efficient. Otherwise it may be necessary to deepen the rafters with battens to get sufficient depth of insulation in, and this may reduce headroom. Grants were available and may still be for some people. Ask the local authority for advice.

### Floors

Extensions provide the opportunity to insert insulation boards between a concrete floor slab and the screed, or under the slab on a concrete blinding. Suspended timber floors at ground level are common in older houses and need to have cross ventilation. However, if access is gained during alterations, laying insulation below the boards should be considered.

### External walls

Where the interior or exterior walls have to be replastered, plasterboard which is bonded to insulation gives higher thermal efficiency, or foil-backed plasterboard with insulation between the battens is an alternative. Modern plasters have more inherent insulation than traditional types, but are less suitable for exposed solid walls, and may not provide a high enough standard of insulation.

### Windows and doors

Double glazing is not suitable for traditional sash windows because the sections are only thick enough to accommodate one sheet of glass. Secondary glazing, i.e. fitting a glazed frame on the inside of the window frame, is one option but this can make the use of the window for ventilation rather awkward. It is effective as a sound insulator, though, particularly if the gap between the existing and the new glass is more than about 75 mm. Proprietary double glazing schemes tend to be expensive, considering that the windows of an average house usually account for about 10 per cent of heat loss, and so only savings of about £20–£30 per annum should be expected. Draught sealing of windows and doors, including the letter plate, can be more cost-effective.

New windows should be double or triple glazed with sealed units. In fact, double glazed units with 'low emissivity' coatings such as Pilkington's Insulight HP have a very low 'U' value (1.9 as opposed to 2.9 Wm² K for the uncoated type). They work on the principle that shortwave solar heat is allowed to pass through the glass but longwave internal heat is reflected back inside the building. Coated double glazed units are as efficient as uncoated triple glazed ones.

### Flues

Chimney stacks with flues should be ventilated even if sealed or disused.

If the fireplace has been removed, the opening bricked in and the chimney pot removed, air bricks should be provided in the external stack and in the chimney breast.

If the fireplace is disused partially reduce the flue with masterboard or similar material, or excessive ventilation will cause draughts.

### Note

If a room has a gas fire without a balanced flue (which draws air from outside) or a solid fuel burning appliance, fresh air ventilation must be provided.

In a room with a suspended timber floor the ventilation between the joists can be made available by drilling small holes equivalent to an airbrick in a floor board – near the side of the room where it will not be covered with furniture or a rug.

### References

- Homebase: *How to Insulate and Save Energy: Leaflet no. 35*
- B&Q: *Insulating your Loft*

## 6.8   Space and water heating

Space heating in the nineteenth- and early twentieth-century house was originally by a coal fire in each room. This required storage for coal, often a cellar filled via a chute from the road. Water heating was usually via a boiler in the kitchen or a back boiler to the fireplace in the rear room.

Space heating by individual gas fires in each room and a multipoint instantaneous water heater over the kitchen sink or bath for the hot water supply became the next solution. Electric storage heaters had the disadvantage of being rather bulky and less immediate.

*Current trends*: Gas space and water heating is the most popular type, despite certain disadvantages.

### 6.8.1 *Gas space and water heating*

*Conventional boiler* with balanced flue supplying radiators and hot water via indirect hot water cylinder and cold water storage tank.

| | |
|---|---|
| For: | Ample supply of hot water. |
| | Provides space for drying towels etc. |
| Against: | Can be difficult to locate tanks and cylinder in small houses |
| | Balanced flue may dictate kitchen layout |
| | Not as economical as a combination or condensing boiler |

*Boiler behind gas fire* using existing fireplace and chimney stack supplying radiators and hot water as above.

| | |
|---|---|
| For: | Uses existing flue but it has to be lined |
| Against: | Less compact and tends to dictate use of room |

*Combination or condensing boiler* with balanced flue directly supplying radiators and hot water.

| | |
|---|---|
| For: | Space saving |
| | Energy saving |
| | Economical to run |
| Against: | Hot water flow can be rather slow for bath – but satisfactory for power showers |

#### Note on wet systems

● All the systems that use radiators are susceptible to leaks and air locks requiring the attention of a plumber.

- They also require pipe runs which are usually cut into the top of floor joists, having an adverse effect on their strength.
- Narrow microbore pipes reduce this problem.
- Thermostatic valves are temperamental.
- In hard water areas furring can be a problem.
- If gas boilers are not regularly serviced they can become dangerous.

Ducted warm air systems avoid the problems associated with water but ducts are less flexible and require careful planning.

## 6.8.2  *Electric space and water heating*

Electric storage heaters are the main form of electric space heating. They are now quite compact. The tops have to be kept clear of wet clothes, difficult to ensure with tenants. Wall panel heaters tend to be expensive to run in comparison.

Water heaters run on off peak electricity have to be large for adequate storage capacity.

Solar heating systems are not yet efficient enough to provide full space heating for a typical UK house. They can, though, provide hot water in the summer provided there is enough external space facing approximately south in which to locate the collectors. In the winter and even in dull summer periods, such systems require backing up with conventional gas or electric systems which increases the cost of the whole installation. Solar electric systems which are efficient and affordable are in the development stage and the likelihood is that with increasing demand and financial backing they will become as commonly used in the UK as in other European countries. Chapter 7 includes a further discussion on the use of solar energy in connection with house extensions.

### *A note on tariffs*

Comparisons between gas and electricity often focus on costs. New sources of energy, combined with a greater flexibility in marketing, may alter perceptions.

## 6.9 The disabled user

It is always possible that upgrading a dwelling will include making provisions for its use by disabled people. Obviously the constraints imposed by an existing building can be severe compared to those experienced with a new one. For the alterations to be affordable, therefore, the proposals are usually a compromise. Old buildings which are fairly spacious may adapt quite well to disabled use but the small Victorian house presents more of a problem even if, as is usually the case, the disabled access is confined to the ground floor only.

Part M of the Building Regulations deals with access and facilities for disabled people but it does not apply to dwellings. Since its introduction in the mid-1980s, however, it has been a useful guide for all building types and for that reason we refer to some of its recommendations here. In the first place, the 1992 edition of Part M clarifies the meaning of 'disabled people' as follows:

- Those with an impairment which limits their ability to walk
- Those who use a wheelchair
- Those with impaired hearing or sight

The definitions are important because it is all too easy to think of the disabled as only those who use wheelchairs and to overlook the often simple provisions which can be made for the other users. The essence of the problem is making the building safe to use, and to walk the building in the company of a disabled person is the best way of spotting what needs to be done.

The following notes are based on the requirements of Part M which can be used for guidance in making alterations for disabled people, particularly wheelchair users.

### 6.9.1  Means of access to and into the building

- Try to achieve a 'level' approach, i.e. one at least 1.2 m wide and no steeper than 1 in 20.
- Use ramps if a level approach is not possible. These should

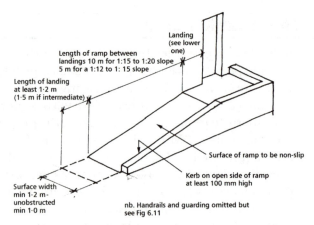

Length of ramp between
landings 10 m for 1:15 to 1:20 slope
5 m for a 1:12 to 1: 15 slope

Landing
(see lower
one)

Length of landing
at least 1·2 m
(1·5 m if intermediate)

Surface of ramp to be non-slip

Kerb on open side of ramp
at least 100 mm high

Surface width
min 1·2 m-
unobstructed
min 1·0 m

nb. Handrails and guarding omitted but
see Fig 6.11

*Figure 6.10    Ramps for disabled people (based on paragraph 1.19 of Approved Document M of the Building Regulations 1992)*

have a non-slip surface and be no steeper than 1 in 15, if the flights are less than 10 m, or no steeper than 1 in 12, if the flights are less than 5 m. See Figure 6.10.

- Use tactile paving slabs on top landings or where the access crosses a carriageway.
- Provide handrails at the side of ramps or stairs.

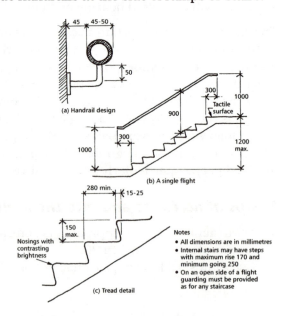

45   45-50

50

(a) Handrail design

300    1000

Tactile
surface

900

300

1000

1200
max.

(b) A single flight

280 min.    15-25

150
max.

Nosings with
contrasting
brightness

**Notes**
- All dimensions are in millimetres
- Internal stairs may have steps with maximum rise 170 and minimum going 250
- On an open side of a flight guarding must be provided as for any staircase

(c) Tread detail

*Figure 6.11    Steps for disabled people (based on paragraph 1.24 of Approved Document M of the Building Regulations 1992)*

- Use dropped kerbs.
- Avoid projections such as outward opening windows.
- Where a stepped approach is necessary (some disabled people find steps easier than ramps) make the tread nosings appear in bright contrast.
- Make steps at least 1 m wide (unobstructed).
- No flight should rise more than 1.2 m – use landings and make these no less than 1.2 m clear in length.
- Steps should have risers no more than 150 mm, goings no more than 280 mm. See Figure 6.11.
- The main entrance door should have a minimum clear opening of 800 mm, have a space of 300 mm to the side of the door's leading edge and have a glazed panel providing vision between heights of 900 and 1500 mm.

## 6.9.2 Means of access within the building

- Internal doors should be similar to the entrance door (above) but the clear opening can be reduced to 750 mm.
- Make internal lobbies large enough for wheelchairs to pass easily from room to room and to turn around.
- Adapt internal stairs to dimensions similar to external (above) except that flights may rise 1.8 m before a landing, the risers should be not more than 170 mm, and the goings not less than 250 mm.
- Make passageways at least 1.2 m wide for wheelchair users, 1 m elsewhere.

Assessing the suitability of an existing house for use by disabled people may reveal other provisions which can be more easily made within a limited budget, such as:

- Fitting door and window furniture (handles, bolts, catches etc.) which is easy to use. Door closers needed for fire purposes should not be too strong.
- Rehanging a bathroom door on the outside to give more space inside.
- Getting rid of single steps, upstanding thresholds etc.
- Improving lighting and fitting vision panels in doors.
- Raising electrical power points to a reachable level.

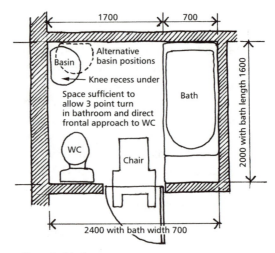

*Figure 6.12   A bathroom for a disabled person*

- Using pull switches for light fittings.
- Lowering fixtures such as shelves and cupboards to a reachable level.
- Fitting handrails wherever they seem to be necessary.

Because Part M of the Building Regulations deals only with non-domestic buildings, it omits any reference to domestic bathrooms and kitchens. These are important because they can present considerable hazards for disabled people. What needs to be done will depend on the degree of disability of the user, so ideally the design should be for one or two particular persons. Among the most important considerations are those listed below.

### 6.9.3   Bathrooms for wheelchair users

- The size and shape of the room is critical. See Figure 6.12.
- The position and shape of the fittings must be carefully considered so that fittings can be utilized effectively. The relationship of person, chair and fitting must be studied. Wash basins, for example, must have knee recesses and the taps must be easily reached. WCs should be approachable from the front or either side.
- Getting in and out of baths is particularly difficult for disabled people. It should be possible for someone to sit on

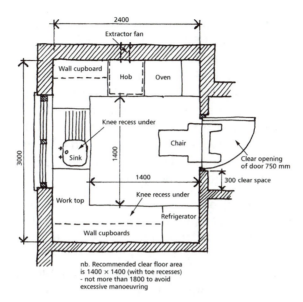

*Figure 6.13   A kitchen for a disabled person*

the edge of the bath and lift the legs over. High-sided baths are therefore preferred. Platforms may be necessary and support rails must be fitted in exactly the right positions. The taps should be lever rather than screw operated.

- Simply raising the WC seat is an effective but simple operation.
- A non-slip floor is essential.

## 6.9.4   Kitchens for wheelchair users

- A cul-de-sac plan is preferred because it reduces possible interference. See Figure 6.13.
- A window over the sink is good for seeing out but an extractor fan will reduce the need for the window to be opened too often.
- The layout of the fittings, their position, shape and size need careful consideration preferably with a particular user in mind. Worktops for wheelchair users should be 800 mm rather than 900 mm high and it is useful to have adjustable pull-out boards at a lower level. Toe recesses should be at

least 200 mm high and 150 mm deep. Knee recesses should be at least 650 mm high but slightly less may be acceptable under a sink.

● Recommended dimensions such as these can be achieved by adapting standard kitchen fittings. Where the budget allows and certainly if a user's needs are likely to change, the use of an adjustable system should be considered. This enables worktops and cupboards, for example, to be fixed at different levels at different times. Flexible pipes enable the sink to be moved as well.

Advice on providing for disabled people can be obtained from the Disabled Living Foundation, 380–384 Harrow Road, London W9 2HU, tel: 0171 289 6111. Locally it is advisable to consult the local authority Social Services department, particularly with respect to grants which may be available for certain disabled people. The Housing Grants, Construction and Regeneration Act 1998 has changed the situation (see Appendix F).

# Further reading

Building Research Establishment (1993) *Double glazing for heat and sound insulation*, Digest no 379. BRE.

Building Research Establishment (1990) *Energy efficiency in dwellings*, Digest no 355. BRE.

Coburn, A., Hughes, R., Pomonis, A. and Spence, R. (1995) *Technical Principles of Building for Safety*. Intermediate Technology.

Goldsmith, S. (1984) *Design for the disabled* (3rd Ed). RIBA.

Johnson, A. (1991) *How to restore and improve your Victorian house*. David & Charles.

Lander, H. (1982) *House and Cottage Interiors*. Acanthus.

Lander, H. (1992) *The House Restorer's Guide*. David & Charles.

Lawrence, M. (1996) *The Which? Book of Home Improvements*. Which? Books.

Williams, A.R. (1995) *A practical guide to alterations and extensions*. Spon.

# 7

# Extensions

This chapter is concerned principally with extensions to:

- Victorian terraced houses
- Semi-detached houses and bungalows of the 1930s
- Individual detached houses

Most attention is given to the first group because a large number still need updating and this can often be combined economically with building an extension.

Extensions fall broadly into three categories:

1 Extending accommodation within the existing shell, i.e.

> Cellars
> Attached outhouses
> Attics

2 Extending beyond the shell as permitted development:

> Roof extensions
> Conservatories
> Ground floor extensions

3 Extensions which require planning permission because they exceed the permitted development allowances.

## 7.1 Extensions within the existing shell

### 7.1.1 Cellars

Traditionally used for storage for fuel, food, wine etc. Now often used for family dining kitchen, particularly when natural light is available. Other possible uses:

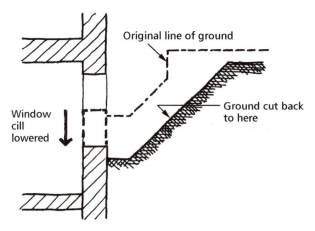

Figure 7.1   Increasing the size of a cellar window

- Spare bedroom
- Table tennis
- Jacuzzi, exercise room etc.

Daylight is usually restricted or non-existent and it may be possible to improve this by cutting back the ground in front of a high-level window as in Figure 7.1. However, windows are not mandatory.

The main considerations are:

- Ventilation
- Adequate headroom
- Damp proofing
- Pumping waste to drains

### Ventilation

The capacity of mechanical ventilation required depends on the volume of space to be ventilated. Consult a services engineer or a manufacturer of ventilation equipment. DIY retailers usually stock a range of fans and ductwork for domestic installations.

### Cellars

Cellars sometimes have inadequate headroom – there are no fixed criteria but 2 m is a useful minimum guideline. To achieve this it may be necessary to excavate but it is essential not to undermine the foundations of the house in the process. Note

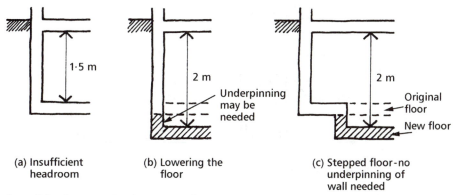

| (a) Insufficient headroom | (b) Lowering the floor | (c) Stepped floor-no underpinning of wall needed |

*Figure 7.2    Increasing headroom in a cellar*

the requirements of the Party Wall etc. Act 1996 discussed in Part 5.1.2. The local building control officer should be asked to state his or her requirements. It is sometimes possible to step the excavation as shown in Figure 7.2.

### Damp proofing

As we saw in Section 2.4, old cellars are often damp. Dealing with excessive moisture is a job for a specialist contractor who will use some of the methods described in Part 2.6.3.

### Pumping waste to drains

An underground location is not as straightforward as a gravity flow to the drains, but waste can be pumped to the required level. A compact macerator pump such as that made by Saniflo Ltd can be used to elevate the waste from a bathroom, including a WC, through small bore pipes (22 mm) to a soil stack at ground floor level. These systems are widely available.

## 7.1.2   Attached outhouses

- The original need for an outhouse, such as fuel storage or an external WC, may have gone, so incorporating the outhouse via an internal doorway is a way of enlarging the kitchen or providing a dining alcove, laundry etc. See Figure 7.3.
- Attached outhouses are often not very well built but provided they are part of the original house, or were built before 1 July 1948, they count as part of its volume for planning purposes. An example is shown in Figure 7.4.

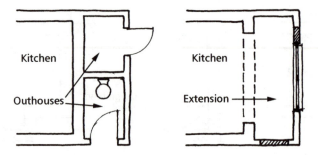

*Figure 7.3   Making use of outhouses*

## 7.1.3  Attics

The key factor in utilizing roof space is that the Building Regulations (Part B: Fire safety) come into force when a second storey is made habitable. This means that:

● The staircase has to be in a fire-protected enclosure and this may have implications for the floor plans.
● Doors from rooms into the stairwell have to be fire doors (half-hour) and fitted with door closers.

Traditional roofs are pitched with the maximum height available at the ridge in the centre of the building. Therefore staircases have to emerge close to the ridge in order to give adequate headroom, unless the roof is extended. There are several ways of achieving this.

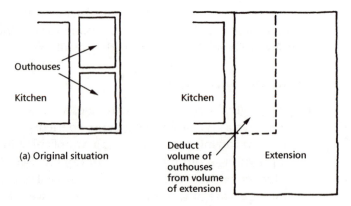

(a) Original situation

Deduct volume of outhouses from volume of extension

(b) Extension replaces outhouses

*Figure 7.4   Extension size*

### Straight stairs above

If the staircase rises from the rear towards the centre, adding an extra flight is simple; it follows the line of the one below as in Figure 7.5.

### Turned staircase

Usually the staircase rises from the front of the house, so a fairly straightforward alteration is to turn the staircase round, to rise from the rear. The staircase then has to have a fire-protected lobby on the ground floor. Although any change to a building should comply with the current building regulations, existing staircases rarely do, but building control officers usually accept turning an existing one. See Figure 7.6.

### Semi-winding staircase in centre

Another solution is a semi-winding staircase in the centre. There is usually a landing of approximately 1200 mm at first floor level above the foot of the existing stairs. This can provide space for

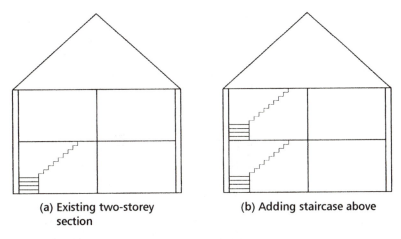

(a) Existing two-storey section

(b) Adding staircase above

Notes:
- Ceiling joists must be checked and probably strengthened or replaced to act as floor joists
- Doors to hall must be fire doors or if existing at least self-closing
- The attic roof must have thermal insulation and ventilation (as discussed in section 6.7)

*Figure 7.5 Access to attic: straight stairs above*

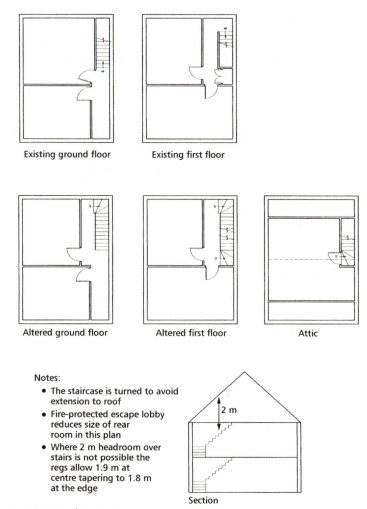

Existing ground floor    Existing first floor

Altered ground floor    Altered first floor    Attic

Notes:

- The staircase is turned to avoid extension to roof
- Fire-protected escape lobby reduces size of rear room in this plan
- Where 2 m headroom over stairs is not possible the regs allow 1.9 m at centre tapering to 1.8 m at the edge

2 m

Section

*Figure 7.6    Access to attic: turned staircase*

a staircase to rise and turn. The straight part of the flight will intrude a little into the first floor bedroom, but the space underneath could become a cupboard, or be metal lathed and plastered to form an elegant curve. The main advantage of this approach is that the character of the existing staircase can be carried right up the house, producing a relatively spacious and elegant appearance. And a rooflight at the head of the stairs can diffuse light widely. The disadvantage is that the staircase is more complicated and so more costly to construct. See Figure 7.7.

### *Semi-winding staircase off centre*

An adaptation of this staircase may be possible over the existing stairwell – however, in most houses it would require a dormer for adequate headroom, and so should be considered as part of the following section or sections, depending on the availability of the permitted development allowance. See Figure 7.8.

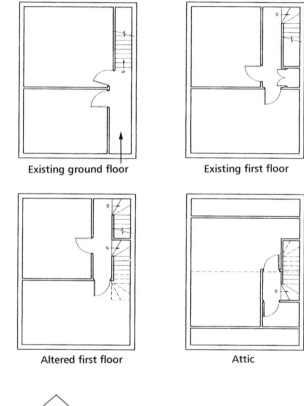

Existing ground floor       Existing first floor

Altered first floor       Attic

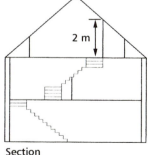

Section

Notes:

- Part winding takes some headroom from main bedroom but room for cupboard underneath
- Small landing at top of stairs gives sense of space and room for chest of drawers etc
- En-suite bathroom, dormer windows or roof lights can be incorporated as required
- Note concession on stairs headroom as in Figure 7.6

*Figure 7.7   Access to attic: part-winding staircase*

### Lowering the ceiling

Sometimes the depth of the house and the pitch of the roof are insufficient to provide adequate headroom in the attic. To keep within the existing shell, one solution is to lower the first floor ceiling as shown in Figure 7.9. There is now no requirement under the Building Regulations for minimum room heights other than the convenience of the occupants. Victorian room heights are often considerably higher than those in new houses or old cottages. Thus lowering the height of a first floor room is one way of adding headroom in the attic. First floor ceiling joists are usually inadequate for floor joists and would have to be replaced by larger timbers or strengthened in any case so lowering the first floor ceiling is a feasible option. On a note of caution, however, it would be most unwise to remove ceiling joists if they are acting as ties in the roof structure without providing stability in other ways. The advice of a structural engineer may be needed and account taken of the Building Control Officer's opinion. It may be necessary to trim the joists and adapt the ceiling around existing window heads for the first floor bedrooms. This is not normally difficult.

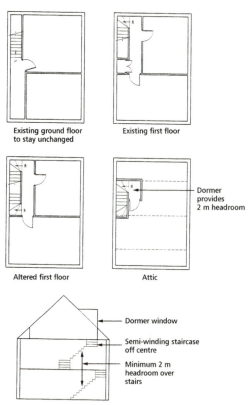

Figure 7.8   Access to attic requiring dormer

### Further notes on attic rooms and attic stairs

If an attic staircase serves only one room (plus a bathroom) it can be 650 mm wide, i.e. 200 mm narrower than a normal domestic staircase. This correspondingly reduces the size of the landing and the area over which 2 m headroom is required. It

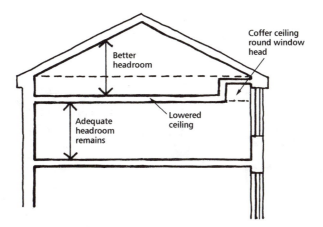

*Figure 7.9   Lowering a ceiling to give more attic space*

is easier to fit a smaller staircase into tight spaces. Whereas 2 m headroom above stairs is desirable, particularly to facilitate furniture moving, it is permissible to reduce this to 1.9 m at the centre tapering to 1.8m at the edge in loft conversions only.

Limiting the number of rooms in an attic has another advantage in that it makes the most of restricted headroom as illustrated in Figure 7.10. People cannot make use of the full headroom if there is a central partition, unless the roof is raised. As an emergency measure Part B of the Building Regulations permits the use of escape windows in the roof of loft conversions. Each room should have an openable window or rooflight in such a position that people can help occupants to escape using a ladder from outside. Clearly this is only feasible where there is space and access below for the use of a long enough ladder. It should not be assumed that only the fire service will make a rescue. The location of the window should be in accordance with Figure 7.11.

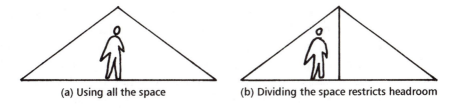

(a) Using all the space          (b) Dividing the space restricts headroom

*Figure 7.10   Making the most of restricted headroom*

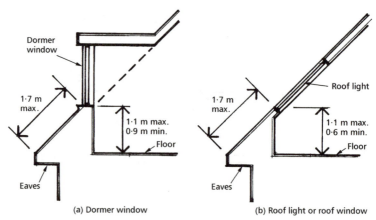

Figure 7.11   Position of an escape window in a loft conversion (from Building Regulations 1992 Approved Document B)

## 7.2   Extensions beyond the shell within the parameters of permitted development

The first point to note is that the permitted development allowance is limited and could have already been used. Once the limit has been reached, any extension requires planning permission. The limits were outlined in Section 5.3.

The following notes are based on the assumption that 50 m³ is the overall allowance for a terraced house, but for a roof extension alone the allowance is reduced to 40 m³. This is for any type of house and cannot be on a side fronting the highway.

### 7.2.1   Roof extensions

Attics in roofs often have ample floor space but much of it is hardly usable because of limited headroom. Raising the roof at the rear increases the use of the attic as illustrated in Figure 7.12. An extension must not exceed the height of the existing ridge. Extensions at the front alter the appearance of a house to such an extent that planning permission is required. Dormer windows are a means of providing localized headroom and can be designed to blend well with most houses. They can also provide a means of escape as we saw in Figure 7.11. They have the added advantage that they are easier to insulate and curtain than rooflights but at the front of the dwelling they require planning permission.

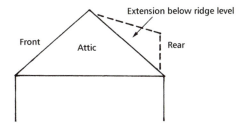

*Figure 7.12 A roof extension within permitted development*

## Roof extensions for detached houses

Provided there is 6 m to the boundary, first floor rear extensions within the permitted development allowance are acceptable, but they must not exceed the height of the existing roof. In the case of a detached house the allowance increases to 70 m³ or more, dependent upon the size of the house (except in conservation areas – see Section 5.3). This enables a two-storey extension to be built, provided that there are more than 2 m to the boundary at the side.

## 7.2.2 Conservatories

Conservatories can be built as permitted development provided that the height of the walls, the apex of the roof and the overall volume are within the prescribed limits. However, they do not have to comply with the Building Regulations. The suppliers of conservatories give guidance on foundations, lighting, ventilation and blinds etc. They have a small but quite useful role to play in collecting solar energy, as we shall see later in the chapter.

## 7.2.3 Ground floor extensions to terraced houses as permitted development

The following options are considered here.

### Enlarge an existing rear extension

See Figure 7.13. The term 'rear extension' is rather confusing. As we saw in Section 6.2, the majority of terraced houses have an L-shaped plan where a 'leg' protrudes from the main part of the house and is generally referred to as an extension. However, in most cases it is an integral part of the house, not a later addition. In any event, for planning purposes, provided it was in existence before 1947 it forms part of the original house.

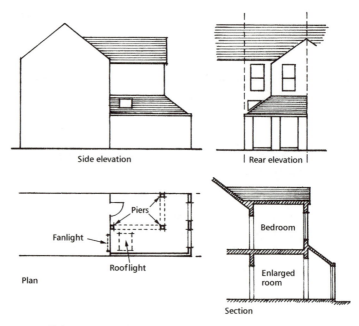

Side elevation

Rear elevation

Plan

Piers

Fanlight

Rooflight

Bedroom

Enlarged room

Section

Notes:

*Figure 7.13 Ground floor extensions for terraced houses: Option (a): enlarge existing back room*

- Room may be 6 m long × width of house
- Piers and beams are needed to support the first floor structure
- Pier sizes may vary - check foundation suitability
- Piers can be a feature of the enlarged room
- Pitched roof minimizes obstruction for neighbours
- A fan light and a rooflight give direct light to the inner room
- French windows or patio doors can lead to garden

### Add a new room

See Figure 7.14. This is a new building and may require very little alteration to the existing house.

### Build a garden room

See Figure 7.15. This is a totally new building and may only require extensions to services such as drainage, heating and lighting.

## 7.3 Notes on permitted development – possibilities for all house types

In reviewing the legislation and practice relating to permitted development, a number of interesting possibilities emerge, including:

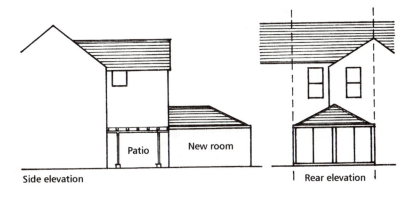

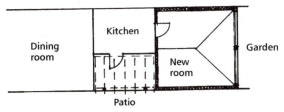

Plan - new room 4·5 m square = 20 sq.m. × 2·5 m high = 50 cu. m.

Notes:

- The roof may be flat or pitched
- A flat roof allows the maximum volume of usable space in the extension but may be visually unacceptable and lead to maintenance problems
- The disposal of rainwater from additional roofs may require the construction of an additional soakaway or alternatively a connection to an existing surface water drain
- Building new walls astride a boundary will require notice to be given to the neighbouring owner as required in the Party Wall etc. Act 1996 (see part 5-1-2)
- Gutters overhanging neighbouring land should be avoided unless by agreement with the owner

*Figure 7.14    Ground floor extensions for terraced houses: Option (b): add a new room*

- Separating new development from the house
- Ground floor extension as the first stage
- Splitting the permitted development allowance
- Adding permitted development before a change of use
- The five-year rule
- Retrospective permission
- Using the permitted development allowance to facilitate change

The essential points of these are briefly noted below.

Plan

Notes:
  • New room must not occupy more than half the garden
  • It may not be more than 3 m high (4 m to apex)
  • The notes on Figure 7.14 apply here also

*Figure 7.15    Ground floor extensions for terraced houses: Option (c): build a garden room*

### Separating new development from the house

Up to half the site of the original dwelling can be covered with buildings ancillary to the use of the house. On a small site this might be less than the permitted development allowance but on a large site it could be quite a considerable amount of building. However, it cannot be:

  • used as a separate dwelling;
  • more than 3 m high (4 m to apex);
  • joined or close to the original building.

But it could be ideal for a games room, a study or occasional entertaining. There is always the possibility of applying for a link later.

### Ground floor extension as the first stage

All the allowance could be used for a flat-roofed ground floor extension as a first stage, on the basis that planning permission for an upper storey would be applied for later. There is a risk that this permission would not be granted; however, pitched roof constructions are generally preferred to flat, so the risk might be worth taking. In any event a flat-roofed building makes the most of the volume allowance.

### Splitting the permitted development allowance

The allowance can be divided between different parts of the house. For example, part could go to raising the roof at the rear and part towards a conservatory. The extension shown as Option (a) in Figure 7.13 for a terraced house is an economical use of the allowance and gives an ample room with several cubic metres left for a roof extension. The allowance can be used in stages.

### Adding an extension before applying for a change of use

Permitted development only applies to single dwelling houses, so if any change of use is contemplated it is wise to consider whether an extra room is needed and build any extensions before applying for a change of use, for example to two dwellings. But bear in mind that planning permission for the change might not be granted, so only add what is necessary.

### The five-year rule

If an extension or even a new dwelling is in existence for more than four years without being challenged by the local planning authority, then it acquires the right to exist, except in the case of listed buildings.

### Retrospective permission

Apart from listed buildings, planning permission can be applied for and granted retrospectively. This is a useful provision if there has been a misunderstanding as to whether planning permission was needed. However, unless accommodation is needed in a hurry, for example to accommodate a bereaved relative, it would be unwise to assume permission would be forthcoming as a result of any informal discussion. Clearly there is a high risk that it may be refused and that the building might have to be demolished, so the whole project and type of construction should be considered carefully. A demountable building could always be relocated.

### Use the permitted development allowance to facilitate later change

It is possible to change the character of a house by altering features such as doors and windows, and to enlarge it considerably without planning permission.

If any substantial changes are planned in the character and size of a house, it is worth considering starting the process as permitted development because there is a tendency for planners to want extensions to replicate the style of the existing, and the size of a new extension is usually considered as proportionate to the size of the existing dwelling.

However, in the case of terraced houses it is important to manage any changes with sensitivity or the terrace could be harmed and the property lose value.

## 7.4 Extensions that do require planning permission

All extensions, however small, require planning permission if the permitted development allowance has been used. First floor extensions to the rear of terraced houses generally require planning permission, because they are usually within 2 m of a boundary. Large two-storey extensions, extensions above 3 m² at the front of a house and extensions which involve a change of use all require planning permission. An example of the type of extension which requires planning permission follows in the part on individual detached houses. Only one (or two) examples can be given because there is less uniformity in this category and the task of referring to more would be beyond the scope of this book. For porches the reader should consult *Planning: A Guide for Households*, generally available from local authority offices.

## 7.5 Other house types

### 7.5.1 Semi-detached houses of the 1930s

An extension within the existing shell may be:

- Utilizing roof space for extra space
- Converting a garage to a reception room for an extra room

#### Extra room in the roof

The considerations concerning permitted development, fire escape provision, staircase, insulation, ventilation and strengthening the structure are similar to those for terraced houses.

#### Extra space in the roof

Sometimes the requirement is for larger rooms rather than more rooms. One way of achieving this is to construct a platform in the roof, reached by a bunk-type ladder from the room below. The platform can be used for a sleeping area, or just for storage, and is ideal for a teenager's bedroom. The roof structure may have to be strengthened and insulated, and sufficient sound insulation incorporated. See Figure 7.16. Access could be from a bedroom on either side of the house, scissors style.

### Converting a garage attached to the house to a room

In most cases this does not require planning permission, but there should be alternative car standing space. Garages are often unattractive and a carport can be an alternative improvement.

The thermal insulation standards will need to be improved and if later building above is planned then the foundations must be checked and if necessary reconstructed for the additional load which such an extension would impose.

## 7.5.2  Bungalows of the 1930s

These are very popular with developers because many are built on large plots.

'Eating bungalows' has become the phrase to describe certain development practices. Bungalows may be 'eaten' by being:

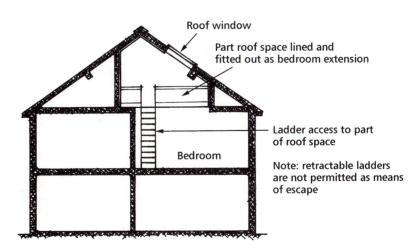

*Figure 7.16  Providing extra space for a bedroom*

- Demolished to make way for two or more houses. This frequently happens when the original standard of construction was poor and subsequent maintenance inadequate.
- Absorbed into larger dwellings.

There are interesting examples of small bungalows being enclosed by larger houses. See Figure 7.17.

Most bungalows cover more ground than the equivalent multi-storey house and to extend them laterally can lead to boundary problems. On the other hand, they often have large roofs which lend themselves to conversions using dormers or roof windows. Beware, though, of the following:

- A lightweight structure designed originally perhaps as a result of poor soil conditions. Such structures should not be overloaded, particularly with concentrated loads.
- Foundations which may not be capable of sustaining the load of an additional floor. Investigation by trial holes will be necessary with the possible result that underpinning will be required, as discussed in Section 2.7.

Except in the case of listed buildings and conservation areas, planning permission is not required for demolition, but planning permission is required for the operations described above. 'Boat house eating' is the term sometimes given for constructing living accommodation over an old boathouse – a very popular adaptation with the increased leisure use of rivers and canals.

### 7.5.3  Individual detached houses

Large extension for a small house: This is usually a difficult proposition because planners think in terms of a percentage of the existing, even though the plot may be large, and accretion over the years may have allowed much larger extensions to

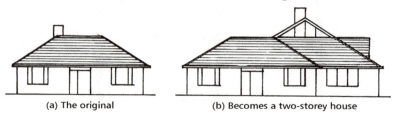

(a) The original          (b) Becomes a two-storey house

Figure 7.17  An extension to a bungalow

emerge in the neighbourhood. The best tactic may be to go for more rather than less than is required, on the basis that this will absorb opposition and the authorities will then be more amenable to a somewhat reduced scheme. Another approach is to add incrementally, starting with the permitted development allowance, which gives more freedom with design and materials, and then applying for more in stages.

### 7.5.4 *Alterations to new houses*

An important consideration in buying a new house is that no VAT is payable. Obviously it is more cost-efficient to buy a new house which meets individual requirements than to alter one with subsequent VAT implications. In some cases this can be avoided by buying on the basis of the developer's plans and agreeing minor changes at the outset. Most house builders will not allow any modifications, certainly not unless contracts are exchanged, because they create extra expense and complications. The expense, of course, would have to be borne by the customer, and might seem rather high, but if it is possible to negotiate this, it is much better than having changes made later. An example would be a town house with a large bathroom on the first floor with a rather restricted living room. Changing the bathroom to a shower room may enable the size of the living room to be increased.

## 7.6 The sustainable extension

An extension or fairly extensive alterations to a house should afford a good opportunity to incorporate energy-saving measures in the scheme. Not only will these bring considerable financial benefits but they can satisfy a need for a healthier lifestyle and one which responds to environmental problems such as resource depletion and pollution. Improvements such as double glazing, higher levels of thermal insulation and efficient boilers have already been mentioned. Each of these can be worthwhile but to achieve the maximum possible effect a whole package of measures should be considered. Whether any one or all can be used will depend on the nature of the alterations and the available budget. Saving the costs-in-use of energy can require a considerable capital outlay in the first place, so a long-term view is necessary. More important perhaps is the owner's desire to contribute something, however small, to the universal effort to conserve energy and protect an environment at risk. This is an area of growing technology but in the UK in the early part of 1998 the following appear to be the most useful options.

## 7.6.1 Materials

- Use materials and components with low embodied energy, i.e. the energy used to produce and transport them to the site. Local sources reduce transport needs and give local character but high technology may have to be imported because the benefits of using it outweigh the costs of obtaining it. It is claimed that softwood requires half the energy of aluminium or a third of the energy of PVCu to produce.

- Use recycled (and recyclable) materials. With care and at little cost a surprisingly large amount of building materials can be used more than once: bricks (particularly if old and built with lime mortar), blocks, tiles, slates, concrete (as rubble), timber, doors and windows etc. Too much of value finds its way into skips. Look out for components which are made from recyclable materials such as glass, most metals but only some plastics.

- Use materials from sustainable sources. Tropical deforestation has led to an enormous reaction against the use of imported hardwoods. Timber should be obtained from certified sustainable sources. Some contractors have joined the Worldwide Fund for Nature's 1995 club which means they will only use certified timber even though it means paying higher prices. In the UK the Good Wood Seal of Approval is given by Friends of the Earth to companies who do not use unsustainable sources.

- Use materials which are user-friendly, such as timber which is lighter in weight and easier to work and handle than concrete and steel, natural products which do not contain volatile organic compounds and water-based paints which are comfortable in use. Some boards which are a good substitute for natural timber, such as chipboard, plywood and MDF board, are out of favour because they produce a formaldehyde vapour, particularly when new. They should be well 'aired' before use.

## 7.6.2 Form and orientation

- Create simple building forms which are economic in the use of materials and components, as recommended previously in Part 3.3.3.
- Design alterations or extensions to make the best use of 'passive' solar gain. Facing south is a simple start but consider the shading effect of trees and nearby buildings at all times of the year.
- Design alterations or extensions to make it possible either now or later to use 'active' solar heating. This usually means creating a roof shape which can accommodate several solar panels facing south (see below).

## 7.6.3 Water

- Devise a system for the collection and use of rain water. Collection is a matter of positioning single or multiple inter-connected butts which must have overflows discharging into surface water drains. Kits are available from garden centres. Rain water can be used for a variety of purposes apart from watering gardens – car washing, window cleaning, topping up ponds and other high-consumption activities.
- Use low-consumption fixtures such as showers, spray taps and low flush WCs. At the same time make fittings accessible and easy to maintain so that leaks cannot go undetected. Ensure that overflow pipes are in places where their functioning will be immediately noticed. Encourage users to use washing machines and dishwashers sparingly.
- Consider grey-water recycling – the waste from baths, basins and washing machines can be stored, treated and either returned for use in WCs or used with rain water for other purposes.

## 7.6.4 Controls

- Use control systems to provide comfort but avoid waste. Modern heating systems are controlled by thermostats and these can be used on boilers, radiators, storage tanks and in the spaces to be heated. Ideally it should be possible to zone the heating so that unoccupied spaces are kept just warm and time clocks can cut off the heating when it is not needed at all.

● Consider the use of 'intelligent' systems where the use of external sensors can enable internal conditions to be adjusted to compensate for the external environment. These may be rather sophisticated for domestic use but serve to illustrate the increasing scope for advanced technology to support the cause of energy conservation.

### 7.6.5  Solar power

In the UK, experiments with solar power began in the years just after the Second World War but it was not until the 1970s that the worldwide energy crisis led to a surge of interest in the subject. In 1996 it was estimated that over 40 000 solar systems were in use in the UK with a further 1000 to 2000 new systems being installed each year. It is not always appreciated that solar energy is derived from daylight, not just sunlight, so its potential for use in the UK is considerable in spite of a rather slower uptake than is apparent in other Northern European countries.

For domestic applications there are three main methods by which solar gain can be utilized. Within the scope of this book it is only possible to comment briefly on these. The reader is strongly recommended to consult the bodies stated at the end for further advice.

#### Passive systems

Buildings which stand in the sun will gain heat, particularly if they are dark in colour. Passive systems utilize the fact that glass transmits shortwave radiation but does not transmit the longwave radiation which is produced by objects such as walls which are heated by the sun so that warm air is trapped between wall and glass. Previously, in Section 6.7, we noted that special coated glass can be used to enhance this effect but two other methods are worth noting. One is the use of conservatories and the other the use of 'trombe' walls both located on the south side of the dwelling. See Figure 7.18. Careful design can utilize the trombe effect by ducting warm air into the house in the winter and cool air into the conservatory in the summer.

### Active systems

These are commonly used to heat water up to a certain level and may also provide space heating. Basic systems have been used to heat swimming pools for many years. They are installed by specialists and comprise a plumbing system including a pump, a heat exchanger, storage tank and solar collectors facing the sun. Figure 7.19 shows a typical installation for an average-sized dwelling. The collectors will be roof-mounted, between three and six square metres in area (for a family of three or four) and tilted at an angle of about 35° to the horizontal. They need only face within about 30° either side of due south. In the UK, solar heating alone can rarely produce water sufficiently hot for many domestic purposes (65°F) so it is usually combined with a conventional system. There is also a need for a control system which reduces heat loss from the collectors and in winter a need to combat frost by using an anti-freeze solution in the primary circuit. 'Evacuated tube' collectors are more efficient than 'flat plate' collectors because each absorber plate is enclosed in a vacuum. They are considerably more expensive, though. Such systems can provide up to half of a family's requirements for

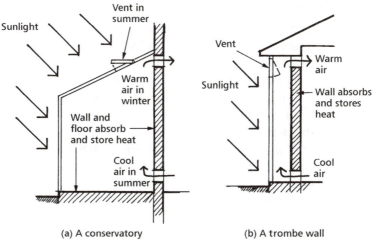

(a) A conservatory          (b) A trombe wall

*Figure 7.18   Conservatories and a trombe wall*

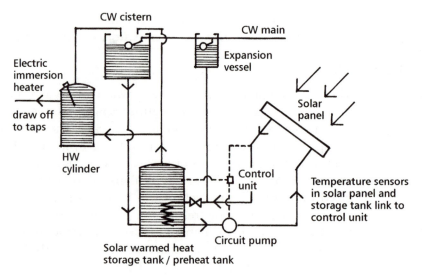

*Figure 7.19   Solar water heating*

hot water, depending of course on demand and the size of collectors used. Most supply between 1000 kW and 1500 kW per year, generally in the summer and autumn, resulting in savings of between £20 and £110 per year. The cost of installation varies from about £1000 to £2500 for a DIY job or £2000 to £4500 for a professional installation.

## Solar electric systems

These are based on photovoltaics which is the direct conversion of solar energy into electricity. Solar cells were first developed in 1954 and are now used to power spacecraft, pocket calculators and a wide range of remote facilities. Their use in buildings has been negligible in the UK but in other countries solar electric systems are becoming widely used. In this case the solar collectors are panels comprising 30 to 50 solar cells which produce a direct current (DC) which must then be converted by an inverter to an alternating current (AC) for domestic use. Surplus electricity can be exported to the national grid. See Figure 7.20.

Greenpeace says that a 2 kW system will produce 1500 units of electricity a year which should cover 40–50 per cent of the average family's requirements. This system will require solar panels covering an area of 12 to 50 square metres depending on the type of panels used. It would have cost approximately £10 000 to install in 1996. This is the equivalent of generating electricity at about 30 pence per unit. Even though exporting to the national grid can bring a return, at about 2.5 pence per unit, this still represents a high cost for producing electricity. Until costs come down it is difficult to see how solar electric systems can be viable for the average householder. However, the following points emerge:

- An increasing demand from those with an environmental conscience and further developments in technology are showing signs that costs will indeed fall in due course.
- The relative costs of alternative forms of energy will most probably increase in time. The steady price of fuels experienced in the UK in the late 1990s is unlikely to continue.
- In due course the UK government, aware of its need for a 'green' image, can be expected to promote solar power more actively.
- Solar panels may be fitted over an existing roof on a metal frame or can be built on to the structure as a replacement for slates or tiles. They are relatively light in weight and only if they represent a net addition to the weight of the roof should it be necessary to check that the roof is strong enough to carry the additional load. Solar water heating will take up some roof space but generally these systems lend themselves to incorporation in an alteration scheme or extension to a dwelling of reasonable size.

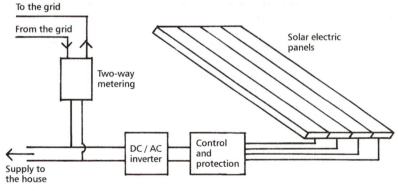

*Figure 7.20   A solar electric system*

### Contact

New and Renewable Energy
Enquiries Bureau
ETSU
Harwell
Didcot
Oxon OX11 0RA
Tel: 01235 432450

The Solar Trade Association
Pengillan
Lerryn
Lostwithiel
Cornwall PL22 0QE
Tel: 01208 873518

Greenpeace Solar Team
Canonbury Villas
London N1 2PN
Tel: 0171 865 8100

UK Solar Energy Society (ISES-UK)
Centre for Alternative Technology (CAT)
Machynlleth
Powys SY20 9AZ
Tel: 01654 702992

# Further reading

Counsell, S. (1990) *The Good Wood Guide*. Friends of the Earth.
CIRIA (1995) *The client's guide to greener construction*. CIRIA.
ETSU for the Department of Trade and Industry (1993) *Active Solar Heating-Technology Status Report 005*. DTI.
Greenpeace (1996) *Building homes with solar power*. Greenpeace.
Kruger, A. (1991) *H is for ecoHome*. Gaia.
Lawson, B. (1997) *Building Materials, Energy and the Environment*. Royal Australian Institute of Architects.
Strong, S. (1993) *The Solar Electric House*. Sustainability Press.

# Buying, selling and finance

## 8.1 Buying and selling houses

In 1997 the whole process of buying and selling houses in England and Wales came under review by the government. There were also moves to introduce a bond payable to buyers to cover survey and legal costs if they were 'gazumped'.

At the time, gazumping was rife because in parts of the country, particularly London, property prices were accelerating. Gazumping occurs in a falling market. Gazumping is the term for the action of a vendor who pulls out of a sale after a price has been agreed because they have received a higher offer. Gazundering means that the purchasers lower their offer at the last minute.

Bargaining often occurs in negotiations before a price is fixed, but the process of buying and selling houses is sometimes very drawn out and the stage at which the price is fixed and binding is quite late in the process. In the meantime the buyer may have incurred considerable survey and legal costs and possibly financial costs as well. However, there may be sound reasons for raising the price or lowering an offer. The sale may take so long that the vendor loses an opportunity to buy at the same level or a survey may reveal unexpected problems for the purchaser. In average market conditions the system operating in early 1998 has many advantages for both buyers and sellers. Buyers do not need to pay for a survey until their offer is accepted and sellers have time to look for another property. Both parties usually regard the offer price as fixed with only minor changes permissible.

The system in Scotland works by means of sealed bids, which once accepted on a set date are binding. The disadvantages are the need to do a thorough survey before making an offer, the offer may not be accepted, and the value of competitive offers is unknown. Repeating surveys for other houses becomes costly and there is no margin for second thoughts over a very expensive purchase.

There is a move now towards setting up 'property centres', common in Scotland, to give the full range of professional advice to purchasers, i.e. financial, legal, structural, etc. The Law Society of England and Wales has voted in favour of this so they should soon become common across the UK, although the position in Northern Ireland is unclear.

### 8.1.1 Conveyancing

The legal process of transferring the ownership of land and buildings from one person to another, i.e. buying and selling, is called conveyancing.

Although it is a process that can be managed by the layman with the aid of a good guidebook such as *Bradshaw's Guide to House Buying, Selling and Conveyancing for All*, unfamiliarity with the process may make it stressful, and there are pitfalls.

Missing or misunderstanding important points may lead to problems which are difficult to put right later, so most people take the precaution of employing a solicitor whose firm would have experience and the backing of professional indemnity insurance.

Where a mortgage is involved the lender would usually insist on a solicitor handling the conveyance and in the case of a complicated leasehold it is particularly advisable to have good professional advice.

Some mortgagors have a panel of solicitors whose work they will accept; otherwise they insist on using their own solicitor which can lead to extra costs and delays. In the meantime a desirable property could be lost. So it is wise to check first whether a solicitor is on the mortgagor's panel. This solution might also apply to the choice of a surveyor.

Even when a solicitor is engaged, it is important for the client to know what is involved because most of the enquiries are referred for an answer to the client by the solicitor. Moreover, the solicitor may take a more cautious approach than a client thinks necessary, so consultation would be helpful. In either case Bradshaw's guide is an interesting account of the process.

A brief outline of the conveyancing process follows. It is not intended to be a DIY guide to buying and selling, which would be more thorough, but will show what usually takes place at the different stages in the process. The notes which follow add some points relevant to this book.

## 8.1.2 *An outline of the buying process for a registered house*

1 Agree a purchase price (having made an offer perhaps).
2 Arrange a firm mortgage if needed.
3 Carry out a survey.
4 Receive a draft contract from the vendor/solicitor and office copies of the Land Register, title and plan and Authority to inspect the Land Register. Investigate the title.
5 Make searches of the local authority about factors which may affect or harm the property, such as road widening schemes.
6 Obtain answers to Enquiries before Contract and any additional information relating to planning permissions or building regulation approvals, for example.
7 Make a bankruptcy search from Land Charges.
8 If satisfactory check property is vacant and co-ordinate mortgage.
9 Exchange contracts with the vendor and agree completion date (usually three weeks later). Insure property. Pay deposit.
10 Make a final search with Land Registry to ensure no new charges have been taken out and to reserve the property for registration (a protection lasting 14 days only).
11 Check that the property is vacant at the time of completion and that nothing unscheduled has been removed.

12 Pay the balance of the purchase price by telegraphic transfer, bank to bank, and receive Land Certificate, Transfer form, Certificate to say that the previous mortgage has been discharged and the keys.

13 Pay the Stamp Duty to the Inland Revenue.

14 Send the Land Certificate, Transfer form, Mortgage Discharge Certificate and proof of payment of Stamp Duty together with the Application for Registration in New Ownership to the Land Registry with the appropriate registration fee.

15 Receive new Land Certificate and a note of Completion of a Registration from the Land Registry.

Most of the transactions are done on standard forms which are obtainable from OYEZ, the law stationers, at Oyez House, 7 Spa Road, London SE16 3QQ. A list of these is given in Appendix D.

### 8.1.3 Notes on the buying process

#### Making an offer

There is no need to wait for a property to be advertised; developers often approach an owner directly or employ an agent to search for them. More usually houses are advertised by an estate agent or vendor with an asking price. In times of flux this becomes a guide price or 'Offers In the Region Of' (OIRO) or more hopefully 'Offers In Excess Of' (OIEO). Valuation is largely a matter of local comparison, although asking prices do not give much idea of what is finally paid – the length of time taken to sell may give a clue.

It is becoming more common to value residential property in terms of price per square metre or square foot. This is in line with overseas practice and can be a useful guideline. However, the number of rooms, circulation space and facilities as well as general state and attractiveness can modify the figure. So can the locality. Paying a deposit to an agent is no longer done. It does not bind anybody and can cause confusion in final accounts.

#### Conditions on offers

An offer to buy is always made subject to contract; this is because until investigations about the title and searches are complete there are too many unknown factors to proceed safely. Subject to survey is usually added and occasionally subject to planning approval. Subject to obtaining a satisfactory mortgage is becoming increasingly common.

*Subject to survey* means that the offer may be altered if the survey reveals anything unexpected or unduly expensive to rectify. However, the vendors may feel that they have taken repairs into account in the asking price and may not be prepared to accept an offer on this basis. But buyers are generally unwilling to go to the expense of a survey unless their offer is accepted, so in most cases it is a standard condition.

*Subject to planning approval* might be tried if the viability of the scheme depends on it, but obtaining planning permission can be a slow process so it might not be accepted. An offer to meet any increase in inflation during the period might make an offer acceptable but it is more usual to enter into a conditional contract with the vendor, and it would be advisable to engage experienced professional help for this. As far as possible it is advisable to work within the parameters of permitted development (see Section 5.3).

*Subject to mortgage offer* – subject to obtaining a satisfactory mortgage offer is a useful provision. It may be that a 95 per cent mortgage offer has been made in principle, but in the opinion of the building society's surveyor the house chosen is overvalued. So 95 per cent of the surveyor's valuation does not amount to 95 per cent of the asking price for the property. If a prospective buyer cannot find the difference, the purchase cannot go ahead at that figure, or a more realistic price might be negotiated.

### Sealed bids

Offers in the form of sealed bids are often required by institutional vendors; they are not subject to survey. For the prospective buyers they can involve a great deal of work in planning and costing repairs and alterations with an uncertain chance of success. Walking round the job with a good builder can be very helpful. The main consideration is not to get caught up in the situation and overbid; there is no point in winning if the enterprise ends in a loss. The highest offer does not necessarily win; some evidence of finance is required and the soundness of this is important. Other factors such as the continuity of work for agents may have a bearing.

### Auctions

Auctions are more stringent for purchasers than sealed bids because the auction constitutes an exchange of contracts and the 10 per cent deposit has to be paid immediately. Auctions are usually chosen when the property is difficult to value and the market is rising, but executors who want to be sure of selling quickly or institutional owners such as building societies selling repossessed properties may opt for auctions. In a market trough they are a good way to pick up bargains and there is the advantage of knowing what others are prepared to pay, but auctions then are rare and are mostly executors sales. Sometimes people do not check the current availability of finance and parameters or rules change, and then they cannot complete the purchase as expected. So it is very important to organize finance and work out all costs beforehand.

### Survey

Doing a useful survey depends on experience and a knowledge of up-to-date prices. With the aid of these it is possible to get a good idea of likely expenditure on updating or extending an average-sized house quite quickly. This should be sufficient for making an offer. If precise information is needed before an auction or a sealed bid, a thorough survey and alternative estimates for work should be obtained.

### The draft contract

This states the purchase price, the deposit to be paid to the stakeholder, earliest date of completion and rate of interest for overrunning the agreed completion date. A stakeholder cannot part with the money until completion. The 'office copy' of the property register gives the title or type of ownership, e.g. freehold and absolute, together with any easements, such as passage of drains, or covenants, for example limits to the height of garden walls. Investigating the title means checking this information and making sure it covers everything wanted and needed. Sometimes items are missing from the plan or description which on site ought to be included. Authority to inspect the Land Register means making a search, which discloses any recent charges and reserves priority of registration for 30 working days.

## Searches

A fee is paid to the local authority which makes internal enquiries about whether there are any environmental health, planning or other problems affecting a property. If in a hurry it is possible to go to the council offices and inspect the registers in person, although mortgagors will want formal written replies. Searches are only valid for a short period, but insurance known as 'validation' can extend the life of a local authority search. Insurance is to cover any loss occurring between the date of the last search and exchange or completion.

## Investigating title

This is a matter of ascertaining whether a property is freehold or leasehold etc. and that the vendor is selling as beneficial owner, i.e. that he or she has a right to sell it. The extent of the property, ownership of boundary walls, and the implications of any easement or covenants need to be explored. If a title or ownership problem is found, for example part of the garden seems to be missing from the transfer, then it is advisable to contact a solicitor to find a solution.

## Enquiries

There is a standard list of enquiries before contract but some solicitors prefer to use their own. Most questions relate to what is included in terms of fixtures and fittings. It is becoming more common for cookers, curtains and carpets to be left by the vendor. Probably the most important factor to ascertain is whether the neighbours have caused problems over noise. It is possible to contact the environmental officer of the council over excessive noise and ask if any notices have been served. Boundaries and trees are a frequent source of contention.

### Date of completion

If the purchase is part of a chain of people buying and selling then the whole process can take a long time, limited to the speed of the slowest member, so the suggested period of three weeks is only for single transactions. In cases where it suits both parties, a completion date could be fixed well ahead. Where there is a hurry it could be very soon but is not often less than two weeks.

### Exchange of contracts

A deposit is payable when contracts are exchanged. Traditionally this was 10 per cent of the purchase price (and in auctions it remains so) but it is now more commonly 5 per cent, particularly as many mortgages are for 95 per cent of the purchase price.

### Financial aspects of buying

Sort out finances before looking. Estate agents are often tied to financial institutions that offer mortgages but it may help to shop around. Mortgage advisors fees can be up to £100 per hour and brokers may be more; however, it depends on the size of the mortgage and whether an offer is taken up. Keep commitments as flexible as possible. See Part 8.2.2 on mortgages. The standard rate of mortgage is around $3 \times$ annual income. When mortgages above normal limits are granted there may be an extra lump sum to be paid as a one-off insurance premium to protect the lender (not the borrower).

## 8.1.4 An outline of the selling process

1 Put the property on the market.
2 Accept an offer.

3   Draft the transfer (conveyance) document.
4   Send a contract to the purchaser or purchaser's solicitor.
5   Respond to queries from the purchaser.
6   Produce documents relating to work done on the property, such as planning permissions, guarantees and certificates.
7   List any fittings and items which might appear to be fixtures but are not included in the sale.
8   Exchange contracts (as for buying above).
9   Wait and make no changes to the house or the property as a whole.
10  On completion, hand over the keys and deeds in exchange for the agreed sum of money.
11  Redeem mortgage out of the proceeds of the sale.

## 8.1.5   Notes on the selling process

### Putting the house on the market

Estate agents generally keep track of prices recently obtained in the locality and so are well placed to give advice on asking prices. It is worth talking to several agents and picking the one who has the best reputation or seems most effective rather than the one who suggests the highest figure or the lowest fee. When an estate agent is employed to sell a property the vendor enters into a contract. The terms of contracts vary and it is important to understand them before signing any agreement. A 'sole agency' gives the right to one agent to sell the property, so if the vendor finds a purchaser they do not have to pay the agent's fee. However, giving 'sole selling rights' to an agent means that the vendor has to pay the agent's fee even if the vendor finds a purchaser. If a 'ready willing and able' contract is signed then the agent's fee has to be paid once they find a purchaser who is able to pay the full asking price and is ready to proceed, even if the vendor decides to withdraw from the sale. An estate agent usually offers a lower fee for sole agency but a wider market may be reached by using several agents. The Office of Fair Trading (OFT) produces a very useful booklet on *Using an Estate Agent to buy and sell your home*. It is available free from public libraries or by telephone from the OFT, tel: 0181 398 3405.

Although the success rate is generally low it is worth trying to sell a house privately, especially if it is in a prominent position where a 'For Sale' sign would be widely seen. People moving to a new area tend to drive around and pick a locality before going to an agent. Local people check advertisements in the newspapers, and they all see signboards. Sale boards are the most effective and certainly are the cheapest way of selling, but they need to look professional. There are firms who offer boards and limited advertising at low cost. See also Bradshaw's guide. Newspaper advertising becomes expensive quite quickly and it is difficult for a private individual to avoid idle enquiries. Although estate agent's fees, plus VAT, can reduce profit margins in small developments, most will offer reductions for fairly frequent transactions. There is a great benefit in that the advertising is continuous and often quite widely circulated; buyers know where to look. However, an extra fee might be required for national advertising and large illustrations. Sifting potential purchasers, checking credentials and accompanying visits are very useful and some agents have an invaluable flair for selling. Also, keeping in touch with agents in order to be notified about other properties could be well worth the fees. Do not allow unaccompanied visits. The key might be kept or copied and it is a legal trauma to remove people if they decide to take up residence.

### Accepting an offer

As far as possible, check that financial credentials are sound before accepting. Ask when the buyer would like possession; if it is a long time ahead negotiations may be drawn out unnecessarily, but that might be mutually beneficial. Set a time frame for exchange. Once accepted, monitor progress and exert pressure if necessary. Chains of buyers and sellers can slow progress. Answer queries if possible but it is quite usual not to know all the answers. Produce photocopies of any guarantees or certificates at this stage.

### After exchange

It is important not to alter or remove anything in the house or garden at this stage, unless it has been agreed. In one instance a purchaser forced a substantial reduction in the price after exchange because the vendor had removed a dead tree, on the basis that the purchaser had intended to use it to support climbing plants.

### At completion

Only exchange the keys for the telegraphic transfer of the balance of the agreed purchase price; bankers drafts are now rarely used. If the purchaser does not or cannot complete, the vendor is entitled to keep the deposit and may even claim damages. If the purchaser delays completion then the rate of interest agreed at exchange has to be paid.

### Letting property

There is now more protection for landlords under assured shorthold tenancies and it takes less time to take action against unsatisfactory tenants. This has revitalized the private housing rental sector. There is still an inadequate supply of housing and increased mobility in the labour market has stimulated the demand for houses and flats to let. Demand and hence the rent varies considerably in different regions and even subtly within districts. However, rents do not vary as dramatically as house prices in different parts of the country, possibly because rents are underpinned by the benefits system. So buying for investment income should be considered over quite a wide area, and much will depend on the standard of management available. The suggested rent will also change with the season, spring and early summer are usually the best times to start a tenancy, probably because they coincide with house sales. Before making any moves, discuss rents and management facilities with letting agents. For tax considerations see Section 8.3.

### Furnished or unfurnished?

Under assured shorthold tenancies it is no longer more difficult to repossess unfurnished than furnished property and in general there is not much difference in the rent. Often the landlord's furniture is an encumbrance. Upholstered furniture has to comply with the latest fire safety regulations and must be labelled by the manufacturer to show that it complies. Taking these factors into account, unfurnished lettings are preferable. However, carpets and curtains should be fitted in most cases as a tenant cannot be expected to cope with such large items of additional expenditure.

### Note concerning assured shorthold tenancies

The lease is initially for six months but this is a minimum. The tenancy can continue without renegotiation until either party wishes to give notice, and therefore if an agent has been employed solely to find a suitable tenant the fee could well cover several more months. The landlord has to give two months' but the tenant only one month's notice of ending the tenancy.

### Agent or private letting

Finding tenants through advertisements privately is much easier than selling without an agent, so the main consideration is to follow the correct procedures in setting up the lease and management. Checking references and collecting rent are also items to consider delegating.

### Letting to companies

There is a standard form for company lets and the usual length of time is one year. This may seem ideal; however, it is important that the tenancy is in the name of an individual tenant with the rent guaranteed by the company, and the company itself should be well known and reputable. Otherwise anyone and any number of people from the company could live in the property and the landlord would have little control.

### Holiday lets

Letting a house or flat occasionally while the owners are on holiday does not normally require planning permission but separating part of a house for continual holiday lets usually does. However, the leases of flats in some apartment blocks specifically precludes subletting; that is, officially it is not allowed and though sometimes letting may still occur there is no guarantee that it will continue unchallenged. Short-term lets produce a high rate of return but the frequent changes between occupants need efficient organization. In the early stages at least it is helpful to have the advice and assistance of a well-established holiday letting agent. Classification by the English Tourist Board can be a help but their requirements are very stringent and rather inflexible.

### Standards

Gas and electrical installations have to be checked and approved every year by qualified personnel. The local authority's environmental health officer has the power to inspect any rented premises and to order remedial action.

## 8.2   Finance

This section is about organizing finance for:

- Buying for owner occupation
- Purchase for investment
- Alterations and extensions
- Small development projects

Working out the availability of finance in broad terms at the outset could save abortive work based on outdated assumptions. Personal resources fluctuate and lending parameters change frequently because they are affected by government policy, the state of the housing market and competition among the lending institutions.

### 8.2.1   Sources of finance

The main sources of finance are:

- Existing capital from savings, windfalls, inheritance, disposable assets, gifts etc.
- Loans against assets such as insurance policies, other properties, stocks etc. The losses incurred in selling some assets, such as insurance policies, often make it preferable to borrow against their security, selling only in emergencies.

The terms of some insurance policies allow a loan against the accrued value of a policy at a very favourable rate of interest.

- Private loans depend on personal and sometimes professional contacts. Any agreement should be in writing, ideally with the help of a solicitor, dated and signed by both parties.
- Loans backed by personal guarantees can be arranged through banks. This is easier if both parties belong to the same institution, as often happens in families. Clearly a guarantor's own assets are at risk and could be forfeit, so this should not be undertaken lightly. Share certificates in blue-chip companies, i.e. well known and tradable, lodged with the bank, are an easily negotiable form of security.
- Grants from local authorities, English Heritage etc. See Part 8.2.4.
- Mortgages, i.e. loans against income, prospects and property. See Part 8.2.2 below.
- Unsecured personal loans. These are offered at higher rates of interest and for smaller sums than loans secured against assets by banks etc.

### Type of loan

The purpose of the project will indicate the type of loan, for example:

| | |
|---|---|
| Short-term development or purchase: | Bank loan |
| Long-term owner occupation: | Mortgage from bank or building society |
| Buy-to-let investment: | Banks and some building societies |

### Amount of loan

This may depend on:

- Current income
- Value of assets

- Efficiency of business plan – cash flow etc.
- Own record with lenders
- Policy of lending institution
- National and local trends
- Amount of capital risked by borrower

## 8.2.2 *Mortgages*

Mortgages are loans or debt based on income set against property. Normally they are limited to the value of the property, although cashback schemes and allowances for setting up the house, i.e. items such as furnishing and paving, may be made for new houses and first time buyers.

Incomes tend to rise with expertise and with inflation, even when inflation is low, and house prices rise over the years despite periodic drops, so mortgage lending is a fairly safe business for the lending institutions. In consequence, mortgage interest rates are relatively low compared with those for loans to purchase items such as cars or boats. For example, 7 per cent could be the rate of interest on a mortgage while 12 per cent would be average for a personal loan and 18 per cent for a second-hand car.

Income tax relief at 10 per cent up to £30 000 of a mortgage is available for owner occupiers, this is known as MIRAS or Mortgage Interest Relief At Source and is administered by the lender. It is less favourable than in the past and, because many economists consider that it distorts the housing market and is possibly unfair, it may eventually be abolished.

However, a mortgage remains one of the most secure forms of lending and the relatively low rates make extended mortgages or second charges a good way of obtaining finance for alterations or a new project. For a further loan it is necessary for there to be some equity left in the house, i.e. for the loan not already to be at the maximum possible, or for income and/or the value of the property to rise. A second charge or loan may be made by a different institution and may be at a different rate of interest. It is preferable to use the same institution if possible, to save on legal and possibly survey fees.

### Types of mortgage

There are three main types of mortgage:

- Repayment
- Endowment policy, personal equity plan (PEP) or pension plan linked
- Interest only

### Repayment mortgage

A proportion of the capital owed is repaid with the interest so the capital debt diminishes. However, the proportions of capital and interest vary.

### Endowment mortgage

A life assurance policy equivalent to all or part of the value of the property is taken out by the purchaser. This provides insurance against premature death and accumulation of a capital sum to repay the mortgage at the end of the term. PEP and pension mortgages are based on the same principle as endowment ones and take advantage of various forms of tax relief.

### Interest only

Interest-only mortgages are available from a number of lenders; they make sense if it is planned to pay off the debt when the house is sold. They keep down the annual cost at the expense of a higher final cost. Nationwide allows existing borrowers to transfer to interest-only mortgages when they reach the age of 60 – a welcome gesture for people facing reduced income.

### Notes

While house prices are rising, mortgages are an efficient way of accumulating capital while living in comfort, but although property generally rises over the long term, there can be severe short-term setbacks which are exacerbated by debt.

All of the linked mortgage schemes are intended to provide a sum equivalent to or hopefully considerably more than the sum initially borrowed. However, the outcome depends on the success of the investment institutions. Some savings policies have not lived up to expectations. There is a wide variation in the performance of different funds and the most widely advertised are not necessarily the best. It is important to check with the financial papers and shop around.

Separating savings and insurance is generally considered to be more cost-effective, i.e. buy term insurance and have a separate savings scheme.

### Negative equity

This means that a property is worth less than the money owed on it. Usually prices recover and the problem disappears. But problems arise if an owner wants to sell and has to pay an estate agent and repay the mortgage if they have no other resources. An estate agent is usually paid directly by the vendor's solicitor from the proceeds of a sale but the mortgage lender has first charge on the proceeds. If all is likely to go to the bank, the estate agent cannot undertake the commission. Repayment mortgages which erode the capital debt are less onerous for purchasers in these circumstances.

### Methods of repayment

The main types are:

- Standard variable rate mortgages
- Discounted variable rate
- Fixed rate
- Cashback
- Capped rate
- Non-standard mortgage

### Standard variable rate mortgages

These are traditional mortgages where the rate of interest varies with Bank of England base rates, influenced by government policy. The building societies do not always react immediately to Bank changes because of the cost of implementing frequent changes. Lending rates are referred to in percentage terms, sometimes the total charge is quoted, sometimes the percentage above Bank base rates.

### Discounted variable rates

In these a discount is offered on the standard rate for a fixed period, usually two years, in return for tying the borrower to a fixed period, say three years, at the standard variable rate. Closing the mortgage before this would result in a penalty of several months' interest, although the mortgage can usually be transferred to another property.

### Fixed rate

Fixed rate mortgages are usually available for periods of from two to five years and involve making a guess on the direction of base rates. The longer the period, the higher the rate is likely to be. If base rates rise then picking a fixed rate at the bottom of the rate cycle is very advantageous. Even if base rates fall they have the benefit of reducing uncertainty for a while.

### Cashback

Cashback schemes were devised to help people over the problem of all the incidental expense of buying a house. In return for a lump sum the borrower is tied for a period with the same lender. These schemes are aimed particularly at first time buyers.

### Capped rate

With a capped rate any rise in the interest rate is limited to a pre-fixed percentage but if rates fall the borrower will benefit from the reduction. Again this mortgage product is tied to a fixed term with the same lender.

### Non-standard mortgage

The Bank of Scotland offered no-interest mortgages for a period. They are based on a loan for a limited period in return for a proportion of the increase in the value of the property over the same period. They are aimed at a minority group, generally close to retirement, but are an example of a useful facility in some cases.

### Mortgage brokers

New types of mortgages are constantly being devised and offered and it may be difficult or time consuming to investigate them all; if so, there are advantages in going to a mortgage broker. Also:

- They are specialists and know the full range of products available.
- Some products are not on the market for long and it could pay to act quickly.
- They are able to compare and explain schemes whereas institutions are only able to discuss their own products.

A broker charges a fee for arranging a mortgage (in addition to any fee charged by the lender) or is paid indirectly through any endowment insurance or other policies. However, a good broker, such as John Charcol, will also offer good advice.

Accountants and solicitors who specialize in conveyancing may also be able to suggest the most appropriate mortgage lender for their clients.

### Remortgaging

Low rates are sometimes advertised which makes switching to another lender seem attractive, or there may be the option to get a much larger loan without paying a higher interest rate for the extra tranche of capital. However, changing lenders is as rigorous as buying a house. A new valuation will be needed, legal checks on the title to the property and assurance that any building work done in the interim complies with planning and building regulations – the latter may pose problems halfway through a job.

Obviously all this work will incur fees and charges, so to avoid some of them it is advisable to apply first to the existing lender for a further advance. They will be reluctant to lose a reliable client and so should help. It may help to persist and even write to head office.

### Loans for commercial/residential developments

There is a trend for converting old warehouses and unfashionable office blocks in town and city centres, and also old dock buildings, into residential units. It is difficult to fund these schemes through the normal mortgage market because the original buildings are not residential.

Finance is needed in three stages:

- Initial purchase
- Temporary finance for alterations
- Residential mortgage – either for sale/purchase or to let

Assurance is needed from local estate agents that expected rents/sales can be achieved and the work has to be carefully planned, scheduled and costed. John Charcol, the mortgage brokers, specialize in assembling such financial packages.

### Buy-to-let schemes

Changes in rental legislation and frequent relocation for employment have stimulated activity in the rental market. Together with the resurgence of the property sector, these changes have encouraged interest in the purchase of property for investment and promoted the introduction of new schemes.

General information on buy-to-let schemes is produced by ARLA – the Association of Residential Letting Agents. Their scheme is operated in conjunction with several banks and building societies. The ARLA scheme is good but hemmed in with restrictions, some of which could prove onerous. They include:

- Rates of interest higher than normal householder mortgages – but lower than commercial mortgages.
- 80 per cent is the maximum loan – 75 per cent in some cases.
- The income from the property to be purchased can only be taken into account if a recent three-year history of rental receipts is available, otherwise annual income from other sources should be from 2.5 to 3 × loan, clear of other commitments. Loans will only be considered if a minimum period of three, preferably five, years is envisaged.
- Property must be let and managed by an ARLA member.

### Other schemes

Other banks and building societies offer buy-to-let finance, including smaller societies who do not advertise. Some have the advantage that they will take the future rental income into account and are well disposed towards older borrowers, on the basis that age usually engenders caution. A good mortgage broker would advise.

**Notes**

The difference in rent between furnished and unfurnished accommodation has lessened significantly – the cost of complying with fire regulations for the former has increased.

Agents on average charge between 12 and 20 per cent of the rent for letting and management, which can include finding tenants, checking references and inventories, drawing up agreements (assured shorthold tenancies), collecting deposits and rent, and arranging repairs.

Mortgage interest, maintenance and repair costs, insurance and management charges can all be offset against tax, also gas and electricity if included.

A realistic return from residential property is from 8 to 10 per cent and despite improved legislation this is not risk free. The main consideration is eventual capital gain.

### 8.2.3  Bridging loans etc.

A bridging loan, technically known as an open bridge, is a loan to enable a purchaser to buy another house before selling their existing one. In a rapidly rising market this can be profitable for a purchaser, who will pay less for the new property and get more for the old one than would be the case if the transactions were simultaneous. However, if the property market is stagnant or falling, a house might take a long time to sell, or not achieve the expected price, and a borrower could have problems and default.

The 1989–95 property downturn caused a revision of the guidelines and banks were only prepared to offer bridging loans after contracts had been exchanged on the sale of the existing house. These are known as closed bridges. The process of purchasing a house has now speeded up, partly because of the more comprehensive registration of properties, so long delays between exchange and completion are less common (unless one party deliberately slows proceedings). These factors have limited the use of bridging loans.

### Joint projects

One way of financing small development projects is to share the cost with others. Professional and trade skills can be pooled and payment deferred. The main advantages are that the risk is shared and the whole process can be more interesting.

The problems are:

- It is difficult to co-ordinate people who have the interest and the finance available at the same time.
- Anticipating all the problems and drawing up a satisfactory agreement between partners – assumptions which seem obvious to one person may not be shared by others. Clarify as much as possible at the outset and write it down – to be signed by all.

### Cash-flow statements

In order to present a project to a bank for funding it is necessary to produce a cash flow which shows how much finance is required, the timing of instalments and the anticipated dates for repayments.

### Interviews

Lending institutions need proof of earnings and assets. It can save time to take copies of current bank and building society statements to an interview, plus photocopies or other records of shareholdings and/or other assets. They may also require proof of identity; a driving licence is useful as it confirms the name, age and address of the applicant.

## 8.2.4 Grants

In theory, grants are available for a wide range of repair and adaptation works to existing buildings but in practice the local authorities and English Heritage, who are concerned with listed buildings, now operate under such severe financial constraints that grants which were once obtainable are now hemmed in with virtually prohibitive restrictions.

Most grants are now discretionary, those for the disabled being the exception.

Grants for renovations are now only for people who have lived in a property for more than three years and it is their principal residence. Thereafter, if the improved property is sold within five years the grant has to be repaid. Most grants are means tested and there is a limit on the amount obtainable; they also have to meet the council's standards on practicability.

*Information on grants* is produced by the Department of the Environment in the Housing Key Facts series and leaflets include:

● House renovation grants
● Common parts grant – for premises with flats
● HMO grant – for houses in multiple occupation
● Disability grants – to adapt premises for disabled people
● Home repair grants – for minor repairs (in some cases materials are supplied)
● HEES home energy efficiency schemes

Grants for draught proofing and heat insulation may also be available for pensioners (the work may be managed directly by the local authority).

Grants for soundproofing may be available to counteract the noise nuisance from new road schemes.

Group repair grants are available when the council plan to repair the external fabric of a group of buildings, such as a terrace of houses in a conservation area.

The Construction Act, due out in 1998, changes the grants arrangements. For a review of this, see Appendix F.

Listed buildings – for information on grants from English Heritage (limited to buildings of outstanding interest Grade 1 or 2*), contact the local authority conservation officer.

## 8.3  Tax considerations

These may affect the choice of property, type of development and timing of activities.

### 8.3.1  Applicable taxes

The following data refer to taxes and rates applicable in early 1998.

Five types of tax may affect property:

- Stamp duty
- Council tax
- Income tax
- Capital gains tax – CGT
- Value added tax – VAT

### Stamp duty

- Is a central government tax payable as a proportion of the purchase price of a property.
- Below £60 000 the rate is nil.
- From £60 000 to £120 000 the rate is 1 per cent.
- Above £120 000 the rate is 1.5 per cent.

### Council tax

- Is a local government tax levied for a wide range of public services.
- Domestic properties are banded from A to H. Band A is for dwellings valued at a market value of less than £30 000 on 1 April 1991. Band H is for those above £350 000. Valuations are set by the county branch of the regional valuation office but the tax is collected by the district council and distributed between the county, district, town or parish councils and the police force.

### Note

Properties are not revalued when alterations are made, but they may be when they are sold. If large extensions have been added, the district council will notify the valuation officer who will then consider whether the extended property qualifies for a higher band. It is possible to appeal against any revised banding for a period of six months.

A valuation band suggests a market selling price, so many people were concerned about bands which they thought too low, a disappointed reaction to valuations made during a property downturn after a dramatic peak.

Exemptions from council tax are available while a building

is undergoing structural alterations, is empty or occupied by certain categories of people, such as students.

Discounts are available for single people living alone. This is because the assumption is that dwellings are usually occupied by two people. There are other categories also eligible for discounts.

For further information on discounts and exemptions contact the local authority.

## 8.3.2 Income tax or capital gains

At present the tax threshold on income is lower than it is for capital gains. Annual income becomes liable for taxation when it is over the personal allowance limit, currently £4045, with the important exception of income from letting a room in an owner-occupied house. The room-to-let allowance is currently £4250, i.e. for the 1997/98 tax year. The capital gains allowance is currently £6500 per annum and capital losses can be carried forward indefinitely to set against capital gains.

Profit on renovating houses for immediate resale counts as income and is liable to income tax after deduction of renovation costs, buying and selling costs and a proportion of incidental expenses such as petrol, telephone, stationery etc. – provided the latter are genuinely attributable to the project.

If houses are let after renovation (and it would be unwise to do this without an assured shorthold tenancy of six months' minimum duration) then capital gains tax would be due on any profit from the sale of the house after the deduction of renovation costs.

### Notes

● Keep all the original receipts.
● Interest on money borrowed for such an enterprise can be set against tax, i.e. it is an allowable expense.

Generally there is no capital gains tax liability on a principal residence but a second residence is liable to capital gains tax if it is sold at a profit. However, if part of the house is exclusively used as an office or a separate flat then that part will be liable for CG tax in proportion to the whole after the property is sold. There are also complications if houses have grounds of more than one hectare and it would be wise in such cases to consult experienced accountants before starting the sale process.

In order to take account of inflation, capital gains tax liability is reduced annually by an indexation allowance based on the retail price index (RPI). The figure is available from the tax office.

If living in, renovating and moving house is undertaken frequently, with substantial capital gains, then an income tax liability might be incurred as this would be considered trading. People involved in the building industry are more likely to be checked by the Inland Revenue. But the average time for people to stay in the same house used to be seven years, so in this period of greater mobility of labour this scrutiny should lessen.

### Notes

- It is appropriate to transfer the capital gains allowance to a second home if it becomes the principal residence.
- Mortgage interest rate relief on taxable income is currently allowed on the first £30 000 of a mortgage. It is allowed at source by the lender and is termed MIRAS.

## 8.3.3  VAT

New houses are currently exempt from VAT but it is payable on all the associated services such as solicitors' fees and Land Registry charges. New work to existing buildings used to be exempt from VAT but it is no longer. All bills for alterations and repairs to existing buildings are subject to VAT, with the exception of listed buildings. Alterations to listed buildings are exempt from VAT but repairs are not. (Many conservationists consider that this anomaly encourages work of a more drastic nature to be proposed than would otherwise be the case.)

*Note*

The implications of VAT are considerable because the tax can amount to a substantial sum. If a large extension is required it is important to explore whether purchasing a larger house would be a more realistic alternative. It is unwise to raise the value above the norm for the locality.

### Complications for self-builders of new houses

Any work done by sub-contractors for a new house, such as a heating installation, is free of VAT regardless of whether the sub-contractor's bill includes the hire of equipment etc. However, a self-builder has to pay VAT on tools, the hire of equipment and professional services. Therefore it pays to organize the work so that design fees etc. are incorporated in other costs, such as asking the supplier of the kitchen units to do the kitchen design, or to sub-contract work such as scaffolding where exemption from VAT would offset savings from DIY. All original receipts for materials and work must be kept, photocopies are not acceptable to the VAT office. Only one claim can be made per dwelling, on occupation at the end of a job. However, the end may be drawn out for the self-builder so it is pertinent to consider whether the interest outstanding on unclaimed VAT is likely to amount to more than the VAT paid on late purchases of wallpaper etc. The date of occupation can also spread over several weeks but the authorities are usually sympathetic.

Information for self-builders is available from the local VAT office, a branch of the Customs & Excise Department.

# Further reading

Bradshaw, J. (1989) *Bradshaw's Guide to House Buying, Selling and Coveyancing for All.* Castle Books.

Office of Fair Trading (1996) *Using an Estate Agent to buy or sell your home.* HMSO.

Office of Fair Trading (1996) *Your Mortgage: A Guide to Repayment Methods.* HMSO.

Wilde, P. (1996) *The Which? Guide to Renting and Letting.* Which? Ltd.

Wright, D. (1996) *The Sunday Times Personal Finance Guide to Renting and Letting.* HarperCollins.

Wright, D. (1996) *The Sunday Times Personal Finance Guide to Your Home – How to Buy, Sell and Pay for it* . HarperCollins.

# 9

# Organizing the job

The extent to which it is necessary to organize construction work in a formal way will depend on several factors such as the nature and size of the job and number of people involved. On the one hand, for large complex jobs a building owner will usually appoint agents – an architect, quantity surveyor and so on to manage the project from start to finish. On the other hand, small jobs can often be run quite successfully by owners performing most management functions themselves.

In the formal contract situations, as we shall see, the management role is given to the architect or in some cases the contract administrator (i.e. someone appointed by the employer to perform the same role). The work on site, however, and all production processes associated with it must be managed by the contractor. In fact, some would argue that the contractor is the best person to manage the design process as well as construction and this 'package deal' or 'design and build' arrangement has much to commend it. Many owners, though, prefer to have the support of a professional team, however small, acting as agents and supervising the project as a whole. In particular it would seem that owners seek assurance that control of progress, quality and finance are in what they perceive to be safe hands. That does not mean that the contractor has nothing to contribute to the management of the whole project, including the design phase. The point at which a contractor is brought into a project has been a subject of discussion for many years. Traditional approaches favour a late entry based on competitive tendering and only after the design and budget have been established in detail. More recent developments have encouraged an earlier input of the contractor at the design stage so that he can make a positive contribution at least to cost planning and job organization.

Negotiated contracts go some way in achieving these aims and we shall be discussing these later (see Part 9.3.3).

This chapter highlights some of the important aspects of job organization and recommends certain procedures that owners will either carry out themselves or delegate to others, depending on circumstances. For simplicity a number of assumptions are made, including:

- Building owners change hats – they are 'clients' to the designers, 'employers' to the contractors, 'owners' to a local authority, 'developers' to others.
- 'Design team' can be one person or several but are usually architects, surveyors, engineers or technicians.
- 'Construction team' can be one person or several but always those who contract to build.
- The emphasis is on traditional practice but with some reference to recent innovatory trends such as the introduction of the New Engineering Contract in the mid-1990s.
- The typical project is a small or 'minor works' type in the private sector.
- The contract is of the 'lump sum' type, i.e. the price is agreed before the contract is signed.

## 9.1 Deciding on a completion date

To decide at the outset when a project is to be completed is logical because it enables all the intervening activities to be properly planned. It is, though, often quite difficult. Certain questions need to be answered before a final decision can be made, including the following.

### 9.1.1 Timing

When is occupation required or, if selling or letting, when is the best time to do this? Tradition has it that selling is best timed for spring or early summer but not winter (particularly around Christmas), but these seasonal fluctuations may be cancelled out if there are strong movements in the local or national housing market.

## 9.1.2 How much needed?

What is the level of completion required? Some owners may not want a contractor to do all the work involved. Typical omissions of this kind are:

- Landscaping
- Painting and decorating
- Laying floor finishes
- Fixtures and fittings

These are popular DIY activities which can follow on from the contractor's work and so cause little interference. The position is somewhat more complex where an activity such as DIY plumbing becomes interwoven with the contractor's work and contractual responsibilities are affected. This is discussed later (see Part 9.5.2 on insurance and the introduction to Chapter 10). In these arrangements two general points should be remembered:

1   Contractors cannot be held responsible for work that they do not undertake or for damage to their work which the work of others may cause.
2   It is absolutely essential that the division of works and responsibilities is agreed at the outset, preferably before tenders are prepared, because the contractor's price for the whole job will be influenced by the number of trades involved and the level of management required. These factors are among several generally referred to as 'preliminaries' which we shall discuss in Part 9.3.1.

## 9.1.3 Finance

*When is finance going to be available?* For various reasons most owners benefit from some kind of financial planning before undertaking a building project. The manner in which the contractor is to be paid is crucial and it will vary according to the size of the job and the time taken to complete it. It is a matter to be agreed at the latest by the time that a contract is signed. Standard forms of contract, which are discussed later, include provisions regarding payment which are worth noting even if a standard form is not being used. Some of these are:

- Contractors are not usually paid in advance of their work. Many unsuspecting householders have been persuaded otherwise and suffered at the hands of unscrupulous 'builders'.
- Interim or progress payments are made by the employer monthly, based on a valuation of the works completed to date. Such valuations can be prepared by a quantity surveyor, an architect or a building surveyor or can be by mutual agreement between the employer and contractor (see Part 9.5.2 on valuations).
- Within a reasonable time of completion, a final account must be settled. Where there is a 'defects liability' clause in the contract or 'retention money' to be held back, the final payment will be delayed for from three to six months after completion (see Part 9.5.2 on defects liability and retention). The implications of this for the employer and the contractor are different but may be equally important.

Figure 9.1 shows how for a typical project lasting four months (the so-called 'contract period'), the payments to the contractor span a period of seven months, although the amounts differ widely. The characteristic 'S curve' of the diagram is an indication that both early and late payments are very much less than those when the job is in full flow in the middle. The low payments at the start can be troublesome for a contractor at a time when his level of investment in the project is high, a matter we refer to again in Part 9.3.1.

In the meantime the diagram indicates that for a period of this size and length, the financial commitment for both employer and contractor is more a matter of cash flow than a lump sum payment. It stands to reason, therefore, that such arrangements should be mutually agreeable to both parties and central to the task of project programming.

## 9.2  Planning the project

### 9.2.1  Some reasons for planning

It is not unusual for contractors to prepare works programmes before construction work starts, although they may be reluctant to commit themselves to a programme which is too rigid and

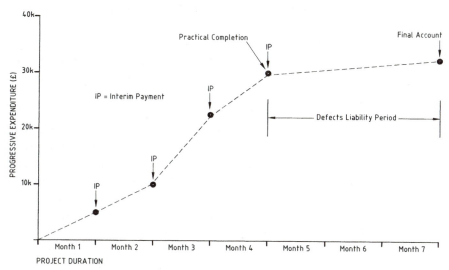

*Figure 9.1    Relationship between employer's expenditure (money paid to contractors) and project duration*

may be used against them in later disagreements. As we shall see later, a contractor may willingly agree to a fixed finishing date but would prefer it if the means of getting there could be left entirely to him. The difficulty with this is that many contractors, who may be excellent builders, are not skilled planners and organizers and, if the project runs into difficulties, they are notoriously inept at taking remedial action. Planning ahead and committing the plan to a works programme can only help, because it provides a clear datum against which all future activities can be monitored.

It is in the owner's interests to see that the whole project, including the design stage, is planned, even to the extent that bar chart programmes are produced. Practitioners and contractors are usually capable of doing this but are not always asked to do so. There is also a general prejudice against producing programmes for small jobs on the basis that the time taken is not cost-effective. Nevertheless, there are benefits to be gained by making the effort, such as:

● It ensures that the sequence of activities is realistic and can be achieved by personnel involved, without over-straining resources.
● It enables all participants to see how their work contributes

or fits into the whole. Start and finish dates are important to those who are only partially involved with the project, such as certain suppliers and sub-contractors.

● It highlights key dates, often deadlines when decisions must be made, materials, colours and fittings must be chosen, or approvals sought, in order to avoid delays.

● It can be used as a progressing tool, keeping a check on whether the work is on schedule and if not, particularly if behind, what can be done in future to restore normal progress.

● It can determine which operations are critical. Without getting into the sophisticated world of Critical Path Analysis, it is usually possible by inspection to identify those activities which are most likely, should their durations be altered, to affect the length of the overall project period.

### 9.2.2 How to go about it

A relatively simple method of producing a programme would be as follows:

● Produce a list of operations in any order, making sure that things which absorb time, such as 'await planning permission', are included as well as those which use other resources, such as 'install new bathroom'.

● Sort the list into an approximate order in which they will occur.

● Estimate the time (duration) that each operation is likely to take. A building owner may well require the advice of a contractor for the construction phase or an architect or surveyor for the design phase. A local authority will always advise on the time required for obtaining approvals.

● Construct a bar chart, using squared paper, showing the anticipated sequence of operations and their durations on an appropriate overall time scale.

Figure 9.2 is a bar chart programme for a typical alteration job showing the main activities included in both the design and production phases.

This programme is based on a number of assumptions which are briefly explained as follows:

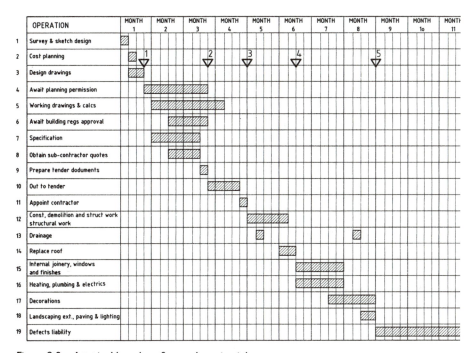

| | OPERATION | MONTH 1 | MONTH 2 | MONTH 3 | MONTH 4 | MONTH 5 | MONTH 6 | MONTH 7 | MONTH 8 | MONTH 9 | MONTH 10 | MONTH 11 |
|---|---|---|---|---|---|---|---|---|---|---|---|---|
| 1 | Survey & sketch design | | | | | | | | | | | |
| 2 | Cost planning | 1 | 2 | 3 | 4 | | 5 | | | | | |
| 3 | Design drawings | | | | | | | | | | | |
| 4 | Await planning permission | | | | | | | | | | | |
| 5 | Working drawings & calcs | | | | | | | | | | | |
| 6 | Await building regs approval | | | | | | | | | | | |
| 7 | Specification | | | | | | | | | | | |
| 8 | Obtain sub-contractor quotes | | | | | | | | | | | |
| 9 | Prepare tender doduments | | | | | | | | | | | |
| 10 | Out to tender | | | | | | | | | | | |
| 11 | Appoint contractor | | | | | | | | | | | |
| 12 | Const, demolition and struct work structural work | | | | | | | | | | | |
| 13 | Drainage | | | | | | | | | | | |
| 14 | Replace roof | | | | | | | | | | | |
| 15 | Internal joinery, windows and finishes | | | | | | | | | | | |
| 16 | Heating, plumbing & electrics | | | | | | | | | | | |
| 17 | Decorations | | | | | | | | | | | |
| 18 | Landscaping ext., paving & lighting | | | | | | | | | | | |
| 19 | Defects liability | | | | | | | | | | | |

*Figure 9.2   A typical bar chart for an alteration job*

- *Operation No. 4.* The statutory period within which the local planning authority is required to make a decision on an application is eight weeks, starting on the date that the appropriate fee is paid.
- *Operation No. 6.* As for planning above, a period of five weeks is the normal maximum for the local authority to deal with a Full Plans application for Building Regulation approval.
- *Operation No. 7.* Writing a specification will not require six weeks of concentrated effort. What the bar shows is the period within which the work must be carried out. This is a general principle applying equally to other operations.
- *Operation Nos 3 and 5.* A distinction is drawn between 'design drawings' which are intended to impress planners and others with the attractiveness of the scheme and 'working drawings' which are for the purposes of costing and construction. Building Regulation approval may be sought using the basic working drawings without details. It is possible, using modern drafting techniques, to adapt design

drawings to create at least part of a set of working drawings which can save valuable time in the early stages.

- *Operation Nos 15 and 16.* The operations which are mostly internal and performed once the building is weatherproof are usually more interwoven and complex than it is possible to show on a simple programme. Not only are more trades involved than the structural work requires, possibly up to four or five, but their work must be phased and inter-dependent. For example, a plumber's and an electrician's works are likely to follow the sequences in Table 9.1.

### Table 9.1   Sequence and integration of trades

| What is done? | Known as? | When? |
| --- | --- | --- |
| Plumber: Fitting main pipe runs and testing them (carcassing) | First fix | After structure but before finishes, linings and floor boarding |
| Electrician: Fixing conduit and trunking | First fix | |
| Plumber: Installing fittings and short pipe runs to same | Second fix | After boards and linings but before decorations |
| Electrician: Pulling through main cables from control units to outlet positions | Second fix | After boards and linings but before decorations |
| Plumber: Add on fittings, commissioning and testing of systems | Third fix | Just before practical completion |
| Electrician: Fixing switch plates, sockets, fittings etc. Commissioning and testing | Third fix | Just before practical completion |

There are many variations on this theme but Table 9.1 gives some indication of the detailed planning required of a contractor once on site. In particular it highlights:

- the need for plumbers and electricians to work at the same time, in confined spaces perhaps, without undue interference or conflict;
- the need for the work in these services trades to be phased in with the work of other trades, particularly carpenters fixing floor boards, plasterers fixing plasterboard, painters, tilers and decorators;

- the need for a DIY owner who elects to do, for example, the electrical work to appreciate the importance of collaboration with other tradespeople;
- the need for someone, usually the contractor's 'person-in-charge', to co-ordinate and control these integrated activities unless chaos is to prevail.

## Key dates (refer to Figure 9.2)

1 *Submit planning application.* The design should be complete before applying for full planning permission. Incomplete designs are best dealt with by applying for 'outline planning permission' as we saw in Part 5.3.7.

2 *Invite tenders.* Before going out to tender it is advisable to have received planning permission and building regulation approval, to have completed the bulk of working drawings and specifications, and to have obtained quotations from sub-contractors or suppliers of special equipment.

3 *Enter into contract.* The period allowed for the appointment of a contractor will depend on how quickly the employer can approve the tender and how much negotiation there is on matters of detail. A contractor may start on site in advance of the contract documents being ready but it is advisable to get them signed as soon as possible. The day on which the contractor takes possession of the site is an important one, not least because it marks the start of the contractor's almost total responsibility for the site, including insurance.

4 *Roof on.* The date on which the roof work is complete and the building reasonably weatherproof (openings awaiting windows and doors should be filled as soon as possible). After this, all the operations that must done in the dry such as joinery, finishes, services and decorations may follow on but in a logical sequence as discussed above.

5 *Practical completion.* The date on which the employer or more usually the architect certifies that the building is complete in that it satisfies all the requirements of the contract documents (see Part 9.3.2) and is fit for occupation. Commonly known as 'hand-over', this is an important date because the responsibility for the building, including insurances, is returned to the owner. Two matters remain

up to the completion of the programme; they are the question of defects liability (see Part 9.5.2) and final payment, both of which occupy, in this example, the last three months of the programme.

# 9.3 Tendering and estimating

## 9.3.1 What do these terms mean?

Tendering is the process by which a contractor submits a price, usually in competition with others, for work fully described in a set of tender documents. For small works at least two, preferably three, contractors should be invited to tender, although the procedures will be much the same if only one is involved. There is a distinction between 'open' and 'selective' tendering. Typically, open tendering is used by local authorities which invite, through newspapers and the technical press, any number of contractors to bid for maintenance and rehabilitation work. This is rarely appropriate for private building contracts. Selective tendering, on the other hand, whereby a short list of approved contractors are invited to tender, remains the preferred method of procurement for most current or prospective building owners.

Estimating is a somewhat ambiguous term. Technically it is the work of a contractor's estimator in calculating how much items of work cost and adding other charges in order to establish a price to be charged. Unfortunately, in the more informal job situation, a price described as 'an estimate' may mean to an employer a firm price but to a contractor an approximate price, and this misunderstanding can only lead to difficulties later on. This situation can be avoided by asking the following questions whenever a 'price, quotation or estimate' is received from a contractor or sub-contractor:

- What information, drawings or specifications have been used as a basis for this price and are they up to date?
- Is this a fixed price or is this person expecting more money if the cost of labour or materials rises?
- When can they start and when will they finish?
- To what extent are alternative methods and therefore costs possible?

● Will the contractor agree that delays or omissions which are of his own making will not give rise to additional costs?

The last of these questions covers the situation, which unfortunately is not uncommon, where the contractor's failure to place orders for materials in good time results in a lack of progress and consequently more expense to him in job overheads such as site management, scaffolding, insurance and other 'preliminaries'. In the more formal situation where a form of contract is used, these matters are properly addressed as we shall see later in this chapter.

A contractor's price for a job is usually based on what the job costs him in terms of:

● Labour
● Materials
● Plant and equipment
● Energy

Having summed these costs, the estimator will add a 'mark up', usually a percentage, to cover profit and overheads. Contractors with low overheads and low profit expectations can submit very keen prices and this has been a feature of tendering practice in times of economic recession. There is a risk, however, that unforeseen costs which cannot be recovered or just inaccurate estimating can lead to financial losses which can seriously harm the contract or the firm. It behoves all those involved with the tendering process to see that procedures are fair and that design information is full and clearly presented so that the risk of inaccurate estimating is reduced.

Contractors are usually asked when tendering to quote 'lump sum' prices and then to give a price breakdown if they are in the running for the job. The breakdown can be based on one of the following documents:

● *Bills of quantities* which have been produced by a quantity surveyor for the main purpose of inviting competitive tenders. We saw in Part 3.2.2 how approximate quantities can be used in cost planning. Full bills of quantities can only be produced when design information is complete, otherwise, as we shall see, they must include an amount of 'provisional items' (see Part 9.3.4). Although they are expensive to produce, they are still considered by many to be the best way of analysing the work content of a building job and of creating a framework for pricing. Table 9.2 gives some indication of the format in which bills are produced. Note that there are four columns: in the first, a letter which with the page number codifies the item; in the second, a description of the item; in the third, the measured quantity; and a fourth, left blank for the contractor to insert his prices.

- A *specification*, usually written by an architect or surveyor, which fully describes the materials and workmanship for each trade involved in the work. A typical extract is shown in Table 9.3. Note the absence of a 'quantity' column. This assumes that the contractor will take off his own quantities in order to arrive at the price to be inserted in the right-hand column.

- A *schedule of works* based on the operations to be carried out, for example room by room or on particular elements, e.g. walls, ceilings, timber frame and so on. We have seen how this can be used for cost planning (in Part 3.2.3) and how with some adjustment and updating the document can be used as a contract document.

A lump sum price can be the basis of a lump sum contract, i.e. one in which the price is agreed before the work starts, as opposed to a 'cost reimbursement' contract in which the contractor is paid the actual cost of the work plus a fee (often a percentage). Small jobs are normally done on a lump sum basis and suitable contracts are discussed in Section 9.5.

### Schedules of rates

Repairs and maintenance contracts can be based on 'schedules of rates'. The tender is based on a specification and a schedule of rates with little use being made of drawings. For example, the contractor may submit a rate for plastering of £8 per square metre. If the job measures out at 100 m² he would be paid £800. The idea is that although the rate can be agreed at the outset, enabling an early start to be made, the precise amount of work can be left to be measured later, perhaps on completion. This means, of course, that the employer must wait for a firm indication of the final cost but has some control over expenditure as the work progresses.

## Table 9.2   A typical page from Bills of Quantities

Section no. 3 – Superstructure
Bill no. 4 – carpentry continued.

| Item | | | | £ | p |
|---|---|---|---|---|---|
| | **Treated sawn softwood continued.** | | | | |
| A | 225 x 50 mm Twice splayed tilting fillet. | 95 | m | | |
| B | 100 x 50 mm Framing to barge board. | 53 | m | | |
| C | 50 x 25 mm Bearers to eaves soffite 170 mm long spiked to sides of rafters. | 182 | No. | | |
| D | 50 x 25 mm Ditto 300 mm long ditto. | 6 | No. | | |
| E | 100 x 50 mm Sprocket pieces 500 mm long built in to hollow wall and spiked to sides of rafters. | 36 | No. | | |
| | **Sawn softwood** | | | | |
| F | 50 x 20 mm packing pieces. | 55 | m | | |
| G | 150 x 20 mm Ditto. | 25 | m | | |
| H | 50 x 50 mm Ditto. | 32 | m | | |
| J | 38 x 38 mm Framing to duct casing. | 280 | m | | |
| K | 38 x 38 mm Ditto fixed to blockwork. | 69 | m | | |
| L | 38 x 38 mm Ditto plugged to concrete. | 19 | m | | |
| M | Extra for short length. | 276 | No. | | |
| | **Treated wrought softwood** | | | | |
| N | 100 x 50 mm Framing in partition plugged to brickwork. | 29 | m | | |
| P | 100 x 19 mm Grounds. | 3 | m | | |
| Q | 100 x 19 mm Grounds plugged to brickwork. | 12 | m | | |
| R | 170 x 30 mm Splayed and grooved fascia. | 95 | m | | |
| S | Fitted ends. | 4 | No. | | |
| T | Mitres. | 4 | No. | | |

Carried to Cost Summary 53.                                                                    £

## Table 9.3   An extract from a specification

£   p

6.03   Joinery – hardwoods and – softwoods shall comply with BS1186: Part 1:1971 and are to be approved by the architect. Exterior quality plywood shall be British made to comply with BS1455: 1972 bonding WBP: Grade 3. Priming of joinery works is to be carried out at the works and to consist of a thick mixture of red or white lead and linseed oil, or priming to comply with BS2521 and 2523: 1966. The joinery manufacturer is to ensure that the moisture contents of the various items of joinery delivered to the site are appropriate to the conditions of use to which the components are to be put. The joiner shall perform all necessary mortising, tenoning, grooving, matching, tonguing, housing, rebating and all other works necessary for correct jointing. He shall also provide all metal plates, screws and other fixings necessary. Where existing windows are being replaced the detailed design of the new windows shall match that of the existing as closely as possible. The joiner shall ensure that all weathering surfaces, throatings, grooves and joints etc and all open connections in external joinery works shall be properly executed, and shall obtain a reasonable degree of weather resistance.

There are standard schedules of rates published, such as that produced by the Property Services Agency (PSA). These speed up the tendering process because the contractor is only required to add or subtract percentages to or from the quoted rates, according to his individual levels of profit and overheads. The complete schedules, though, are rather bulky and cost £200 for the seventh edition in 1995. For a particular project it is preferable for an architect or surveyor to devise a schedule of only those rates which apply to the job in hand. Either way, the rates are easy to check and can be compared with currently published ones. Comparing one tender with another requires skill, however, because contractors will usually submit different combinations of rates higher and lower than the standards.

The principle involved with schedules of rates is worth following even if the schedule is not the only basis for a contract. Included as a tender and contract document, a schedule of rates can be helpful in settling the price of variations and in cost-planning future work. For example, a schedule may show a rate for painting ceilings of £1.60 per square metre. A variation requires an extra 40 square metres to be painted so the cost of the variation is 40 × £1.60 = £64. Some time later, the same rate could be used for a similar job elsewhere with allowance being made for increases in the costs of labour and materials and probably for different site conditions.

### Preliminaries

The costs to the contractor of managing a contract and operating effectively on site are called 'preliminaries' and they are usually itemized in a specification or bills of quantities. As we saw in Part 3.3.1, some preliminaries are time related so vary with the duration of the work. Others are either related to the cost of work or are one-off payments independent of time or cost factors. When tendering, contractors may price preliminaries as separate items, include them all in a lump sum or absorb them in their prices for measured items. In other words, in the example given above where a contractor's rate for plastering was £8 per square metre, we can assume that, say, 80p of this (10 per cent – about average) would cover preliminary charges. If that is not the case, i.e. a quoted rate does not include preliminaries but these are priced separately, then that must be clarified in the document-ation. A contractor must cover his operating costs somewhere.

Some typical preliminaries:

- Mostly time related: site management team, plant hire, site huts and storage, lighting, telephones, safety, security.
- Mostly cost related: water, insurance, miscellaneous charges as a percentage of the contract sum.
- Mostly one-off charges: temporary roads, fences and hoardings, erection of plant, setting out, drying out, clearing up and cleaning.

Competitive tendering attracts a range of bids and the lowest of these will be the favourite to carry out the work. There is a need for caution here, though, and the tender breakdown should be scrutinized for errors and omissions. If an error is discovered, the contractor may be asked, while other bids remain confidential, whether he wishes to correct it or to stand by his price. If he corrects the error, and the result is a higher price, he may lose out to the next highest bidder. If he decides not to, there will be a suspicion that he will try to make good his loss by cutting his costs in other ways. Another reason for checking the lowest tender is to look for prices which are artificially low, evidence that the contractor is desperate to win the contract and will use other means to make it pay.

Yet another practice to be aware of is that of 'front loading'. In this case an estimator will inflate his prices for the early stages of the job, such as excavations and foundations, to improve the firm's cash flow at a time when it would otherwise be rather weak. This was referred to in Part 9.1.3 and is common practice. There is a risk, however, to be borne largely by the employer, that should the contractor fail to complete the contract, the partly completed building will be worth considerably less than the price paid for it. A vigilant quantity surveyor can spot front loading in a set of priced bills and, when excessive, it can lead to the suitability of the contractor being questioned.

A high bid is often an indication that the contractor does not want the work, unless it can be very profitable, but wants to remain in favour rather than letting his client down by dropping out altogether.

On balance it would seem that contractors stand to gain more in the long term by producing accurate estimates rather than distorting them for short-term gains. Putting in low prices in the expectation that by devious means money can be recouped later has proved to be a very risky strategy.

### 9.3.2   *Tender documents and contract documents*

By far the greatest difficulty faced by an estimator who is trying to produce accurate estimates is that of incomplete or misleading information. As we have seen, tender prices must be based on a clearly defined package of information, often described as the 'tender documents'. Typically these are a set of drawings showing the full extent and location of the works together with one of the documents described in Part 9.3.1. These were:

- Bills of quantities
- Specification
- Schedule of works

For tendering purposes, the drawings may not show all details provided they are amply described in one of the other documents. The drawings must not mislead, however, by omitting to show one of the floors, for example. It has always been customary to highlight new work in colour (usually red) in contrast to the existing building, left plain (or coloured blue).

The tender documents may have evolved from those used in the earlier cost planning stage – the schedule of works, for instance – and having been priced and amended by agreement they should form the basis of the contract documents for use as the work proceeds.

Contract documents as described in standard forms of contract are not letters, memoranda or minutes of meetings. They are the documents to be signed by both the employer and the contractor, as parties to the contract, and should be a well-co-ordinated and accurate set of information. Each party should have a signed copy of the documents and the contract and these must be kept securely without alteration. Any alterations made necessary as the works proceed must be made on copies of the documents, not the originals. Any change to the works made after the contract documents have been signed will almost certainly be a variation affecting the contract sum (see Part 9.3.6 on variations).

Typically the Intermediate Form of Building Contract (IFC84) lists the contract documents as:

- contract drawings numbered
- the specification
- the schedules of work
- bills of quantities

delete as appropriate (any one would suffice)

The Agreement for Minor Building Works (MW 80) is different in that bills of quantities are excluded and the use of the word 'schedules' as opposed to 'schedules of work' is taken to mean that 'schedules of rates' may also be used (as discussed earlier).

### 9.3.3 Tender or negotiate?

#### Selective tendering

The traditional practice of obtaining competitive tenders for building work is still popular, particularly with employers, architects and surveyors. Selective tendering by a group of well-chosen contractors (as opposed to open tendering which is rarely appropriate) offers several benefits, including:

- Subject to the provisions discussed above, the lowest tender can be taken as the market price for the job.
- The need for the design team to produce tender documents in a complete package means that the amount of information available for pricing is at least well advanced, if not complete.
- It is a well-tried system which everyone understands.

The main disadvantages of competitive tendering are said to be as follows:

- It is time consuming.
- Selection based on price alone overlooks other important factors such as quality and site control.
- It is assumed that the contractor has nothing to contribute to the design process.
- For every contract the estimator wins by this means, 10 may be lost. Across the industry this is wasteful of resources.

- As we have seen, keen competition encourages dubious pricing and the possibility of conflict over prices later on.

### Negotiation

One of the alternatives to selective competitive tendering is negotiation. In this case only one contractor is selected, possibly after discussions with several, and asked to submit a tender. Unlike other methods, though, the contractor can play a useful role in the pre-tender stage, particularly when his experience in such matters as programming, construction methods, alternative costs and materials can contribute to the design process. As discussed in Section 3.2, the contractor may have already played a considerable part in developing a schedule of works, for example.

The main advantages of negotiation are:

- A contractor who has been given a reasonable assurance that he will be awarded the contract will normally be prepared to disclose his prices in some detail so that alternatives or omissions and additions can be discussed and agreed. This is helpful to a design team that may be working to a tight budget. To be sure that the contractor's prices are fair, it is useful for someone in the design team (probably a quantity surveyor) to be knowledgeable in this area and to have access to current cost data.
- The contractor is selected on the basis of suitability, not cost alone.
- An early start is possible sometimes by 'opening up' parts of the building, partial demolitions or site exploration, all done by the contractor on a 'jobbing' basis while the bulk of the contract is still being discussed. This has the effect of shortening the overall development period.
- The contractor will benefit from a longer than normal 'lead time' so that programmes can be prepared and resources allocated well in advance of starting on site. In particular, materials and components on long delivery can be ordered in good time, thereby reducing the risk of later delays.

The main disadvantage of negotiation is the lack of any assurance that the resulting tender represents a good market price for the job. As we have seen, there is a need for someone 'on the owner's side' to be knowledgeable enough about prices to check that what is agreed is reasonable. Most negotiations are about prices with the contractor giving advice on methods etc. Negotiations on matters of design are less straightforward. Owners may wish to play down the contractor's involvement with design on the basis that it may confuse the issue of responsibility if the design is found to be unsatisfactory later on.

### Two-stage tendering

A third option exists which attempts to combine the advantages of both selective tendering and negotiation. Generally known as 'two-stage tendering', it has become very popular over the past few years for jobs of any size. The two stages are as follows.

*Stage One*: normally up to six contractors are selected. They are asked to submit bids based on schedules of rates, typical measured items or preliminaries which relate to the proposed job but fall somewhat short of a full tender. The contractors are usually also asked to submit proposals on job planning and organization, sub-contract arrangements and so on. Full discussions of the proposals are encouraged by inviting each contractor individually to a meeting with the client and the design team. In this way the client will be able to form a good opinion as to the contractor's suitability for the work, and of course the contractor will benefit from discussing the job with his potential customers.

*Stage Two*: the successful contractor from Stage One can now be involved with the final design of the project, ensuring that all contractual and operational matters receive the benefit of his advice. At the same time, a full set of bills and/or specifications and/or schedules of works are prepared and priced by negotiation with the contractor. This then leads to the signing of a contract, but one advantage of this method, as we saw in negotiation generally, is that the contractor may be asked to carry out specified enabling works (opening up an existing structure, for example) immediately after Stage One and well before Stage Two and the contract documents are complete.

## 9.3.4 Problems with conversions and alterations

Even in new work it is quite impossible to remove all the uncertainties associated with building. For various reasons, the many bodies that tend to become involved may not perform in the way expected of them. Building is manufacturing but the product is usually a one-off, is produced in the open air and sits in ground which it is notoriously difficult to assess beforehand.

Alterations are even more prone to uncertainties, variations and additional costs in consequence. Broadly there are five possible sources of difficulty:

1   The original survey may be deficient because the building is occupied or full of furniture. Carpets and floorboards are not lifted, cupboards and roofspaces not accessible and so on. There is the additional problem, as we have seen, that the building may be listed, which prevents any probing of the fabric below the surface. Dilapidation is a further hazard, particularly if the structure is dangerous and prevents a proper survey being carried out.

2   Even if the survey is thorough, an old building may have been altered in the past in such a way that analysing its structure (how it stands up) is difficult.

3   The owner may be uncertain as to his or her precise requirements, a condition exacerbated perhaps by the complex and doubtful condition of the building. Most owners want to defer making decisions for as long as possible.

4   The design team may misinterpret the owner's requirements or may introduce technical or costing errors at the design stage which, when combined with other uncertainties, can have serious consequences. After all, the construction of a wall is never certain until the plaster is removed.

5   If the building is listed, particular care must be exercised in the design and construction work to meet conservation requirements. There are many judgements to be made and even when experienced people are involved, as should be the case, there are differences of opinion as to what should be preserved, conserved, reconstructed and so on.

Where a particular item of uncertainty can be identified, it may lead to the use by the architect or surveyor in the tender document of a 'provisional sum' of money to cover anticipated expenditure on that item. All an estimator has to do in this case is to include the sum in his tender even though it may be a poor estimate of what the item is likely to cost – that is the responsibility of the architect or surveyor.

Tender documents for conversion or alteration works tend to include more provisional sums than for new build. This is regrettable because even with a lump sum tender an employer will have an unclear picture of the true cost of the job. Also, contractors are often tempted to overprice items for which provisional sums have been allowed to compensate for losses on measured items which they make deliberately to keep their original tender low.

To take a simple example: A measured item of brickwork should be priced at £2000 but the contractor, in order to be competitive, prices it at £1000. It actually costs him £2000 so he loses £1000. The contractor may not be too concerned, though, because the bills include a provisional sum of £3000 for certain demolition work and he is asked to do this before agreeing a price. His charge for the work is £4000 which is at least £1000 higher than it should be but the architect can find no reason to dispute it. The contractor therefore wins back the £1000 he lost on the brickwork item. This rather simple story serves to emphasize the need, when dealing with provisional sums, to obtain and accept quotations before giving instructions to proceed.

Once construction work starts it can be very difficult to hold prices of variations down unless they can be based on firm criteria such as a schedule of rates agreed at the time of tendering. This is why it is useful to have a schedule of rates included in the tender documents and in the contract documents which follow.

### 9.3.5 Prices – fixed or variable?

It is essential, prior to inviting tenders, to decide on what kind of tender is required, particularly whether it should be based on a 'fixed price' or one which allows 'fluctuations'. A fixed price implies that fluctuations in the costs of materials and labour (usually increases) will not be allowed once the contract gets under way. The estimator must anticipate these and make suitable provisions in his tender. The longer the contract, the more likely it is that increases in cost will occur.

Labour costs are influenced by national pay awards made every year and are therefore fairly easy to predict. Material costs, on the other hand, may follow a general trend upwards but are much more susceptible to unexpected movements. Figure 9.3 illustrates the way in which tender prices and building costs have increased over recent years. It can be seen from this that they both closely followed the increase in the Retail Price Index until 1988. Then the national economic decline caused tender prices to plunge while costs continued to rise steadily. It was 1992 before tender prices recovered to their original rate of growth.

Where fluctuations are allowable, the estimator need only base his estimate on current rates, safe in the knowledge that any increases occurring within the contract period will be reimbursed. The contract document will provide for this, as we shall see in Section 9.5.

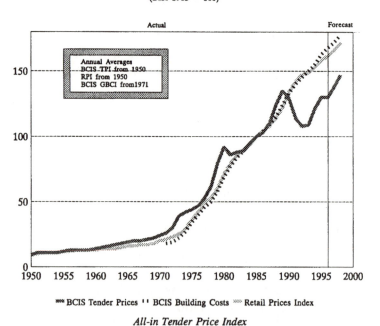

**TENDERS, BUILDING COSTS & RETAIL PRICES**
(Base 1985 = 100)

*All-in Tender Price Index*
*General Building Cost Index*
*Retail Prices Index*

Figure 9.3    *Tender price and building cost indices (from BCIS Indices and Forecasts: Sept. 1997)*

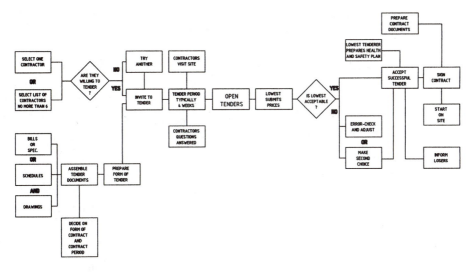

*Figure 9.4   Procedure for single-stage selective tendering*

Even when a fixed price is involved, the very nature of conversion and alteration works, as we have seen, is very likely to lead to variations in the work actually required and this affects the contractor's costs. In some cases a variation will result in a saving, the omission of certain works for instance, but more commonly a variation will result in additional costs. In either case the contract sum (the total amount payable to the contractor) will require adjustment.

Figure 9.4 shows a typical sequence of activities which may comprise a single-stage tendering procedure. It follows a traditional approach and assumes that the tenders are 'lump sum'. Broadly speaking, the sequence can be applied to a negotiated tender (one contractor involved) or selective tendering (several contractors).

## 9.3.6   Variations

As we have seen, the nature of building alteration work is such that variations are extremely likely to occur. They stem from uncertainties at the design stage, unforeseen circumstances as the works progress, late changes in the owner's requirements and so on. Clause 3.6 of the JCT Minor Works form of contract typically states:

The Architect/The Contract Administrator may, without invalidating the contract, order an addition to or omission from or other change in the Works or the order or period in which they are to be carried out and any such instruction shall be valued by the Architect/the Contract Administrator on a fair and reasonable basis, using where relevant prices in the priced Specification/schedules/schedule of rates, and such valuation shall include any direct loss and/or expense incurred by the Contractor due to the regular progress of the Works being affected by compliance with such instruction.

Instead of the valuation referred to above, the price may be agreed between the Architect/the Contract Administrator and the Contractor prior to the Contractor carrying out any such instruction.

As a statement of intent this is admirable but in reality variations give rise to problems such as:

- A complex variation may be difficult to price if it falls outside the prices in the specification or schedule of rates.
- An 'emergency' variation must be executed quickly before the full cost implications are realized and certainly before costs in advance can be calculated.
- Variations can cause delays in contract progress which may be difficult to assess.
- An unscrupulous contractor will allow less for omissions and more for additions, in terms of their true values.
- An unscrupulous employer will resist paying for variations which he will see as someone else's fault, even though they would appear to be justified.
- When variations result in additional costs, savings elsewhere in the contract may be needed. A breakdown of prices based on bills, specifications or schedules will be an invaluable aid to this process of keeping costs under control. On no account should it be left to the contractor alone to 'save a bit here and there'.

## 9.4 Selecting a contractor

For alterations and small-scale building works, the selection of contractors to be included in a list of tenderers, or of a simple contractor in a negotiated contract, will depend on a number of factors including:

- *Size of job.* Approximately 80 per cent of all UK building contracts are within a total cost of £150 000 and carried out by small or medium-sized firms or the 'small works' divisions of larger contractors. Firms to fit a particular size of job are common in most areas, both urban and rural. The fact that a job is relatively small should not deter owners from seeking out suitable builders. The building industry in the UK has always been an industry of small firms. In 1982 the proportion of private contractors employing fewer than eight people was 87 per cent. By 1992 this figure had increased to 94 per cent and these firms did 28 per cent of the total value of new private work but an impressive 47 per cent of all repair and maintenance work (Housing and Construction Statistics 1984–94 (1996) HMSO). Caution is advised, though – this after all is the realm of the 'cowboys' who should be avoided at all costs.
- *Nature of work.* Working on existing buildings requires skills which are in some respects different from those required for new building but no less important. Listed buildings impose particular demands on the contractor, not least because he must be sensitive to the demands of both the building and the conservation officer who will be overseeing the work.
- *Availability.* There is much to be said for employing contractors who are available locally. For small-scale work familiarity with local resources such as labour and plant can be a distinct advantage for the keen contractor. Some experience of working with the local authority and service undertakings will also prove useful.
- *Reputation.* Personal recommendations from people who have employed a contractor in the past can be a useful guide as to his general performance, site management, co-operation and so on. The opinions of more than one person should be sought if possible. Checking a contractor's financial situation is more difficult but again former employers will usually have had some experience related to this.
- *Experience in management of safety.* The CDM Regulations 1994 place particular responsibilities on the contractor to manage and co-ordinate safety procedures on site and to prepare a Health and Safety Plan. Selecting a contractor who

has a poor record in safe building and pleads ignorance of CDM would seem therefore to be unwise.

- *Previous jobs.* The keenly competitive market of the late 1980s and 1990s has forced many contractors to engage in various kinds of marketing activity. Some have produced brochures, others albums of photographs and videos. Most will offer prospective employers the opportunity to visit completed buildings and this, coupled perhaps with personal recommendations, is a very satisfactory way of assessing a contractor's capabilities.

In compiling a list of tenderers it should be remembered that candidates should be equal in terms of their suitability and capability. Ideally there should be no favourites. To initiate the search for suitable contractors it is often helpful to contact one or more of the trade or professional organizations to which most reputable contractors belong. These include the following.

### For general contracting

Building Employers Confederation (BEC)
82 New Cavendish St
London W1M 8AD
Tel: 0171 580 5588

Federation of Master Builders (FMB)
Gordon Fisher House
14–15 Great James St
London WC1N 3DP
Tel: 0171 242 7583

Chartered Institute of Building (CIOB)
Englemere
Kings Ride
Ascot SL5 7TB
Tel: 01344 23355

National House Building Council (NHBC)
Chiltern Avenue
Amersham
Bucks HP6 5AP
Tel: 01494 434477

Guild of Master Craftsmen
166 High St
Lewes
Sussex BN7 1XU
Tel: 01273 478449

### For services

National Inspection Council for Electrical Installation
Contracting (NICEIC)
Vintage House
37 Albert Embankment
London SE1 7UJ
Tel: 0171 582 7746

Heating and Ventilating Contractors' Association (HVCA)
ESCA House
34 Palace Court
Bayswater
London W2 4JG
Tel: 0171 229 2488

Council for the Registration of Gas Installers (CORGI)
St Martin House
140 Tottenham Court Road
London W1P 9LN
Tel: 0171 387 9185

Institute of Plumbing (IOP)
64 Station Lane
Hornchurch
Essex RM12 6NB
Tel: 01708 472791

### Others

National Federation of Roofing Contractors Ltd (NFRC)
24 Weymouth St
London W1N 3FA
Tel: 0171 436 0387

British Decorators Association (BDA)
32 Coton Rd
Nuneaton
Warwickshire CV11 5TW
Tel: 01203 353776

British Woodworking Federation (BWF)
82 New Cavendish St
London W1M 8AD
Tel: 0171 580 5588

# 9.5 Forms of contract

## 9.5.1 Standard forms

The contractor's tender constitutes an offer which the employer may accept or not, and once accepted it forms a binding contract. An exchange of letters may be sufficient, provided that the contractor's offer is made on the basis of, and refers to, information provided by the employer or his agents. We have seen in Part 9.3.2 that this information is normally a specification or schedules and drawings comprising the tender documents.

The terms of the contract are also relevant. 'Implied' terms are those which can be assumed to apply, 'taken for granted'; for example, the contractor must exercise skill and care in carrying out the work, or the employer must provide information as required and so on. 'Express' terms are those in which either party has a stated obligation to act in a particular way, for example the contractor must employ a 'person-in-charge' or the employer must insure against loss or damage to the existing structure (in alteration works). It is the express terms which form the basis of the conditions of a contract and these, coupled with forms of agreement, are published as standard forms of contract. These documents are available for a wide range of types and sizes of construction projects. For small to medium works the following may be used.

### Joint Contracts Tribunal (JCT) Standard Form of Contract for Work of a Jobbing Character

First introduced in 1990, this simple form removes the vagaries and difficulties associated with small jobs procured in a very ad hoc, even casual, way by most building owners. The contract is between the employer and the contractor, no reference being made to consultants. The contract is based on estimates but drawings and specifications are not necessarily used.

### Joint Contracts Tribunal (JCT) Agreement for Minor Building Works (MW80)

Used for small projects of greater value and complexity than jobbing works, this contract allows for the appointment of an architect and the use of drawings and specifications (but not bills of quantities) as the basis of a lump sum price. It has been used for quite large contracts but is best suited to those without complex sub-contracts and which have a value of less than £100 000. This form of contract has been used as a model for the matters discussed in Part 9.5.2.

### The Faculty of Architects and Surveyors (FAS) Form of Contract

This is available in two forms: Small Works and Minor Works. These forms are very similar in content to the JCT ones. Again, drawings and specifications are used but not bills of quantities. Nominated sub-contractors are not referred to and neither is the architect. The word 'advisor' is used instead so that a surveyor may be employed in this role.

### Joint Contracts Tribunal (JCT) Intermediate Form of Building Contract (IFC84)

Structured in the same way as MW80 above, this form is more complex and considered suitable for lump sum contracts up to about £350 000 in value, although because it is easier to understand than its big brothers, the JCT80 forms, it has been used for considerably larger contracts. It suits contracts lasting no more than about 12 months and with relatively simple service installations. Unlike MW80, sub-contractors can be named by the employer. An architect (or contract administrator) is involved and a quantity surveyor is required to value any variations even though bills of quantities are only an optional contract document, as we saw in Part 9.3.2.

### Engineering and Construction Contract (ECC)

This was first published as the New Engineering Contract in 1993 and represents a radical departure from traditional forms. In July 1994 Sir Michael Latham in his report 'Constructing the Team' recommended that it should be adopted as a national standard for almost any construction job. The ECC uses common language because it has to be translatable for overseas use. For the first time the contract is directly related to project management and is intended to stimulate good team collaboration because it defines the individual's functions very well. A project manager must be appointed and a works programme is to be used in joint decision making with the contractor. An adjudicator will settle disputes. For financial control either a bill of quantities or an activity schedule is used, the latter being similar to the schedule of works described in Part 9.3.1. The contract clauses are of two types: 'core' clauses which deal with such matters as communications, the programme, defects, insurances and so on and 'option' clauses from which may be chosen the particular method of procurement and payment mechanisms required. An outstanding feature of the contract is its versatility. The ECC is one of the NEC family of contracts to which will be added in due course a Minor Works variation. At the time of writing (December 1997) this has yet to appear.

These forms of contract are less complex versions of those used for medium to large projects but they follow a fairly general format. The JCT Minor Works Contract (MW80), for example, has the following main headings:

1.0　Intentions of the parties
2.0　Commencement and completion
3.0　Control of the works
4.0　Payment
5.0　Statutory obligations
6.0　Injury, damage and insurance
7.0　Determination (when parties are in default)
8.0　Supplementary memorandum (dealing with VAT etc.)
9.0　Settlement of disputes – Arbitration

In 1995 MW80 was amended to take account of the CDM Regulations. The main additions were under statutory obligations in which the duties of the contractor and principal contractor, whether or not they are one and the same, are described. The essence of these and similar amendments to other standard forms is that compliance with CDM and therefore safe construction has become a contractual matter. As we approach the end of the 1990s, the implications of this for future contracts are just beginning to emerge.

Forms of contract are generally disliked by small contractors who regard them as unnecessarily complex and legalistic. For this reason many owners will be persuaded that merely writing out what needs to be done is an adequate alternative, particularly if such a document can be checked by a solicitor.

Whatever the chosen form, a contract document should include the agreed procedures for the following:

- When work is to start and when it is to finish
- How to deal with delays – reasons for acceptance – reasons for non-acceptance
- Method of payment – monthly valuations – retention fund
- How to deal with variations/price changes
- Who is to insure and against what risks
- How to deal with defective work – during and after completion
- The work done by sub-contractors – contractor's responsibility
- The work to be done by the owner, if any

A simple 'home-made' contract document is given by Mike Lawrence in his book *The Which? Book of Home Improvements* published by Which? Ltd in 1996. This is reproduced in Appendix E.

In the next few pages some of the most important of these factors are discussed in more detail.

### 9.5.2  Some important contractual matters

#### Valuations

As explained in Part 9.1.3, monthly valuations become necessary when the contractor is due to receive progress or interim payments as the job proceeds. These valuations are based on a measurement of work completed on the site by the agreed date and are agreed between the design team (usually the quantity surveyor) and the contractor. In the formal contract situation the valuation is the basis of a certificate prepared by the architect or contract administrator which informs the employer of the amount due to the contractor.

Apart from the works completed on site the standard forms of contract also provide for valuations to include:

> the value of any materials and goods which have been reasonably and properly brought upon the site for the purpose of the works and which are adequately stored and protected ... (Agreement for Minor Building Works MW80)

and in some cases:

> the value of materials and goods before delivery thereof to or adjacent to the works ... set apart at the premises where they have been manufactured ... clearly and visibly marked ... (Standard Form of Building Contract 1980)

This means that the price of a set of doors, for example, which have been ordered by the contractor and are waiting, packaged, at the factory before delivery to the site, can be included in a valuation. The responsibility for ensuring that such goods have been properly ordered and set aside is in the first place the contractor's, but the person who certifies that payment for the goods is due to the contractor is also responsible and should seek confirmation that this is the case or check with the manufacturer. Otherwise there is a risk that the employer is paying for goods which do not exist and may never exist if the manufacturer ceases to operate. In a formal contract the architect, surveyor or contract administrator will probably be aware of the risks involved with valuation and certification. In the absence of any formal procedure, though, the employer would be well advised to scrutinize valuations and seek clarification on what is included and what is not.

### Retention

As a safeguard against the possible failure of a contractor to complete the work or remedy any defects arising from the work, it is now common practice for an employer to withhold a percentage of interim payments until the works are satisfactorily completed. Standard forms of contract suggest a retention of 5 per cent with 2.5 per cent being returned on practical completion and 2.5 per cent when final defects have been made good. It is not unusual, though, for the employer and contractor to negotiate a lower rate. In fact, contractors generally dislike retention for obvious reasons and some may resort to inflating their prices to compensate for the loss of income it causes. Others will accept it grudgingly and appreciate the intention behind it, which is to get the job finished and paid for as soon as possible.

### Defects liability

Under the terms of most building contracts, a defect is a fault in the materials or workmanship for which the contractor is responsible. Clause 2.5 of the Minor Works Agreement 1992 states that:

> Any defects, excessive shrinkages or other faults to the Works which appear within three months ... of the date of practical completion and are due to materials or workmanship not in accordance with the Contract or frost occurring before practical completion shall be made good by the Contractor entirely at his own cost unless the Architect/the Contract Administrator shall otherwise instruct.

The period of three months is referred to as the 'defects liability period', as we saw in Part 9.2.2. Defects which appear in this period should be listed (usually by the architect/surveyor and contractor in a process known as 'snagging') and must be made good by the contractor in a 'reasonable time'. In practice, the defects liability period in most jobs is increased to six months and where service installations are involved a period of 12 months will ensure that they are seen to function effectively through all seasons.

Defects which occur before practical completion, i.e. during the progress of the works, can be dealt with in the way described. Defects which occur after the defects liability period are a different matter. They are referred to as 'latent defects' and the contract cannot be applied in such a situation. If an owner becomes aware of latent defects in a building, he or she may wish to sue those deemed responsible for negligence. However, under the terms of the Latent Damage Act 1986, the action will probably founder if the defects occurred more than six years before the writ is issued, even though the owner may not have been aware that the defect had occurred at the time.

The difficulties experienced by owners in taking action in tort for the alleged negligence of a contractor or designer have led many developers to require 'collateral warranties' to be drawn up by their agents. These warranties are essentially contracts which exist alongside but separate from the main agreement in which the responsibility for defective work and therefore the defendant's liability is considerably greater than would be the case in tort. An example would be that of the designer of a roofing system being required to give a warranty that it will function satisfactorily for a period of 25 years or more.

Understandably the design professions do not care for collateral warranties and there are alternative ways of providing protection. One such is for the employer to take out insurance, such as Building Users' Insurance against Latent Defects (BUILD) which is now more widely available.

Some contractors see the defects liability period as a nuisance, in some cases because it delays their final payment unnecessarily and in others it would seem that the small amount of money involved does not justify the effort of returning to put defects right. In fact, from a legalistic point of view the defects liability period is in the contractor's interest. Without it the contractor would have no legal right to return to the site and the employer could sue for damages if defects became apparent.

### Insurance

In general building work there are three basic risks which should be covered by insurance. They are:

- Injury to or death of persons
- Injury or damage to property
- Loss or damage to the works and site materials

Most reputable contractors take out insurance to cover their liability in these respects but a small operator may not, and because building is such a hazardous business it pays to check what is actually covered by a policy, whether it is held by the employer, the contractor or in their joint names. Most building owners or employers insure their properties, but when alterations are planned the extent of cover afforded by a policy should be checked with the insurance company or their brokers.

A simple contract document for alteration work may state that 'the Contractor will take out appropriate employer's liability insurance and third party liability insurance to cover the work'. This is convenient but far too loose; we are not sure what 'cover the work' may mean and in any case the employer's responsibility to insure the existing building is not mentioned. It is in the standard forms of contract that we find the insurance requirements spelt out in full and often complex terms. The Agreement for Minor Building Works (MW80) provides a relatively simple version of what has become the standard and for this reason the essential requirements are noted here.

### Clause 6.1 Injury to or death of persons

This means any person sustaining any injury 'arising out of, in the course of or caused by' the works unless caused by the negligence of the employer or anyone for whom the employer is responsible. Contractors and sub-contractors must be insured. As far as employees are concerned, the policy must comply with the Employer's Liability (Compulsory Insurance) Act 1969.

### Clause 6.2 Injury or damage to property

This is any property (but not the works) damaged 'arising out of, in the course of, by reason of' the works but only if the contractor or his servants or agents are negligent or guilty of a breach of statutory duty, omission or default. As in Clause 6.1, contractors and sub-contractors must be insured but in this case MW80 requires a sum to be inserted as the minimum cover required. This will vary according to an assessment of the risks involved but a figure of several £million is not unusual. MW80 does not include any provision for damage to property where the contractor is not proved negligent (e.g. the collapse of an adjoining building – a genuine accident) but the larger forms do and special insurance can be taken out by the employer to cover such risks. Neither is damage caused by the employer's negligence covered, but again insurance can be obtained, possibly jointly with the contractor.

### Clause 6.3A Insurance of the Works – Fire, etc. – New Works

This applies to new works and the contractor is required to insure (in the joint names of employer and contractor) against loss or damage by 'fire, lightning, explosion, storm, tempest, flood, bursting or overflowing of water tanks, apparatus or pipes, earthquake, aircraft and other aerial devices or articles dropped therefrom, riot and civil commotion, for the full reinstatement value thereof plus ... % to cover professional fees, all work executed and all unfixed materials and goods delivered to, placed on or adjacent to the Works and intended therefore'. These risks are usually covered by a contractor's 'All Risks Insurance' but the policy may require an endorsement to the effect that it is to be in the joint names of the employer and contractor. The percentage to cover professional fees should be inserted in the clause; it will probably be the same as that which applied to the original job, from 5 to 15 per cent perhaps. Note that this insurance does not cover the contractor's huts, plant or equipment, whether owned or hired.

### Clause 6.3B Insurance of the Works – Fire etc. – Existing Structures

This is the alternative clause to 6.3A, so that one or other of them must be deleted. In this case it is the employer's responsibility when undertaking conversions or alterations to insure the existing building and its contents against the risks listed in 6.3A. Most building owners will have suitable insurance in place but once again it is advisable to check that it is the equivalent of an All Risks policy in the joint names of the employer and the contractor.

### Clause 6.4 Evidence of Insurance

Either party (employer or contractor) can ask for evidence that the appropriate insurances have been taken out. This is very important and both parties would be well advised to check with their insurance brokers if they have any doubt about the validity of their own or the other party's policies.

There would seem to be little opportunity for things to go wrong if the clauses of a typical contract are rigorously applied. In practice, however, problems do arise, including the following:

- Architects and surveyors should be sufficiently knowledgeable to advise their clients on the basic requirements for building insurances. It is, though, a rather complex business. For example, it may not be appreciated that if there is a risk of damage being inflicted on an adjoining property, the special insurance referred to under Clause 6.2 above should be taken out. It must not be assumed that such damage will be the fault of the contractor in which case Clause 6.1 may apply.
- If the contractor fails to maintain adequate insurances the employer may insure and deduct the cost of doing so from monies due to the contractor. If the employer defaults under Clause 6.3, the contractor may insure and add the cost of doing so to his account.
- Sub-contractors are required to 'take out and maintain' insurance against risks to persons (6.1 above) and property (6.2 above) but not the works (6.3 above). The contractor must ensure that sub-contractors hold suitable policies but many experience great difficulty in doing so.
- The employer's negligence (or that of persons for whom he or she is responsible) is specifically excluded from Clause 6.1 and not mentioned in Clauses 6.2 and 6.3. Employers should be aware of this, particularly if they are involved with the works as DIY enthusiasts. Even if they are 'enrolled' as sub-contractors by a well-meaning contractor, they would still need to be suitably insured and should seek advice on a suitable policy from their brokers.

● Contractors' All Risks Insurance policies vary from company to company so it is unwise to assume that all possible eventualities will be covered by them. In particular, certain kinds of hazardous works such as deep excavations, high-level work, demolitions and road transport may be specifically excluded.

### The Housing Grants, Construction and Regeneration Act

Part II of this act, due to become law on 1 May 1998, will have a considerable effect on certain clauses in future forms of contract. Residential occupiers will not be affected but the act will have repercussions on contracts and procedures of all kinds. There is to be a 'scheme for construction contracts' published as a Statutory Instrument, which will make it a legal requirement to follow certain procedures. The arrangements for settling disputes by adjudication are set out and, perhaps more importantly, the methods of making payments are much improved. The issues and practices discussed in this chapter appear, at the time of writing, to be unaffected. For a brief review of the act, see Appendix F.

# Further reading

Anderson, C.A., Miles, D., Neale, R. and Ward, J. (1996) *Site Management Workbook and Handbook.* ILO.

Aqua Group (1996) *Contract Administration.* Blackwell.

Aqua Group (1990) *Tenders and Contracts for Building* (2nd Ed). BSP.

Brook, M. (1993) *Estimating and Tendering for Construction Work.* Butterworth.

Chrystal-Smith, G. (1983) *Practical Guide to Estimating for Alterations and Repairs.* International Thompson.

Clamp, H. (1993) *The Shorter Forms of Building Contract* (3rd Ed). Blackwell.

Franks, J. (1991) *Building Contract Administration and Practice.* Batsford.

Ramus, J. and Birchall, S. (1996) *Contract Practice for Surveyors* (3rd Ed). Laxton.

Smith, A. (1995) *Estimating, Tendering and Bidding for Construction.* Macmillan.

Walker, A. (1996) *Project Management in Construction.* Blackwell.

Willis, C.J. and Willis, J.A. (1991) *Specification Writing* (10th Ed). BSP.

# 10

# DIY activities

There are a number of building operations which are commonly performed in modernizing older houses and which are often attempted and may be managed by the DIY enthusiast. A typical list is considered in this chapter. The notes are intended as a general introduction to the complexity of each activity and to highlight some of the associated problems. They are not a comprehensive or technical guide. It is important to be realistic about the requirements in skill, time and effort involved with DIY work. Skill brings speed and it is often much quicker and so more cost-effective to employ a craftsman rather than to undertake DIY, although DIY for some can be mentally rewarding and interesting.

Enthusiasm is the basic requirement for DIY but it should perhaps be tempered with some regard for the following:

- The safety aspect is very important. Suppliers of equipment and materials are legally bound under the Health and Safety at Work etc. Act 1974 (HASWA 74) to provide instructions as to how to use their products safely. The instructions should be read and understood. Some things are more hazardous than is often realized. If in any doubt at all, telephone the manufacturers who will always provide technical advice.
- Get help when it is needed. As we shall see, all the major DIY chains, for example Homebase, Wickes, Do-It-All and B&Q, produce literature which gives advice on tools, materials and DIY methods. Some also have experienced staff available to answer customers' questions. Bookshops and libraries have DIY manuals and there are a number of relevant periodicals. Be sure that the manuals are up to date

and thus comply with the latest regulations. In seeking out help, do not overlook the hardware shop on the High Street (regrettably diminishing in number) where staff are only too ready to offer advice.

- If the work is concurrent with other contractors' work on site then the question of insurance raised in Part 9.5.2 should be addressed. It makes good sense for the DIY worker to take out a good personal accident policy and to ensure that the contractor's policies cover his or her working on the site.

- As we saw in Part 9.2.2, DIY work may be interwoven with other trades, particularly at the stage of finishes and service installations. Sequences need to be agreed and an overall programme drawn up which the DIY worker should be prepared to work to. This is important because, as we have said, DIY work tends to be slower than one would expect from a skilled tradesman.

- When work of a specialist nature is being carried out jointly with a contractor, it would be advisable to check that the DIY contribution does not invalidate any guarantee for the work.

- For obvious reasons, the quality of workmanship and finish of DIY may not be up to the standards of skilled craftsmen, although there are exceptions. This may be a problem where a contractor has overall responsibility and would not want poor work to reflect badly on the whole job, but it can be resolved by mutual agreement and understanding. It is, after all, quite common practice for the contractor to be asked to leave at least some of the internal decorations and fixtures to the client. As we saw in Part 9.1.2, the division of responsibility can be agreed at the outset so that no confusion arises later as to 'who does what'.

## 10.1   Digging holes and trenches

Many digging jobs can be done by mechanical diggers which can usually be hired from local plant hire companies. Note the following, however:

- The minimum period of hire is usually one day.
- The charge may be reasonable, say £60 for a 1.5 tonne machine for the day, but haulage to and from the site will

be charged extra, say £30 for a 15-mile return journey, and VAT would be payable on these charges in addition.

- Access may not be possible. Excavators can be lifted by crane even over a terrace of houses but the cost would usually be prohibitive. Gateways, even doorways, may not be a problem because 'mini-diggers' can squeeze through quite small openings. A Kubota 1.5 tonne machine, for example, is only one metre wide.
- Steeply sloping ground may prevent some machines, particularly wheeled types (as opposed to tracked), from working safely. Waterlogged ground may be another hazard. Clearly it makes sense to discuss any unusual site conditions with the hire company beforehand rather than leave it until the machine arrives on site.
- Excavators are heavy, awkward machines and very dangerous in untrained hands. Skilled driver/operators are available, however, and should be used wherever competence is in doubt.

Hand digging can be very hard work, particularly where tree roots have to be cut with an axe (beware of causing life-threatening damage, not permissible if a Tree Preservation Order exists) or where rock has to be broken out. In the latter case it may be necessary to hire an electric concrete breaker of the Kango type for about £20 a day (+ VAT). These machines operate on 110 volts electricity so a transformer is needed if using mains supply. Beware, though, both the Kango and the transformer are very heavy indeed.

Whatever the method of digging, the following points should be noted:

- Keep top soil separate. It is far too valuable to be buried below inferior soils.
- Ideally there should be sufficient space at ground level adjacent to the hole for the spoil to be safely stacked. Having to move the spoil more than once (a form of double handling) should be avoided.
- Below a certain depth there is a risk that the sides of the trench or hole will collapse. The risk is greater with sandy, gravelly soils than it is with clays and greater still if the ground nearby is being subjected to additional loads such

as building materials or vehicles or even the wheels of the excavator doing the work. If the soil conditions are known to be poor and the depth of the hole or trench is in excess of about one metre, then some form of 'trench timbering' should be used. Figure 10.1 shows a typical arrangement for a hole 1.2 m deep. Deep holes in poor soils are best left to contractors.

- Ground water, or excessive rain water for that matter, can make digging quite difficult. Water also renders most soils less stable so the risk of collapse is greater. Unless the water is likely to drain harmlessly away it may be necessary to use a pump to keep the hole water-free. A submersible pump will cope with fairly clean water but a diaphragm type may be needed if the water is mucky. Either can be hired for about £25 per day + VAT.

- Whatever the purpose of the hole or trench, it should be backfilled as soon as possible. This should be done in layers of about 200 mm using the original soil (except that in the case of drains it is normal practice to cover the drain with a fine gravel) and compacting each layer. Compacting can usually be done by hand using a stout timber, or even the heels of a pair of heavy boots, taking care not to damage whatever is being buried (a drainpipe, for example). If possible leave the backfill for a week or two for the soil to settle before paving or final levelling of the surface. If a flat

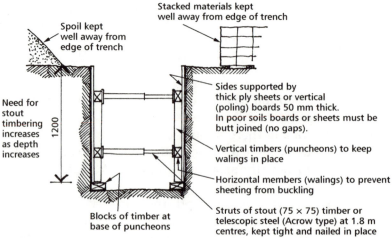

*Figure 10.1    Trench timbering*

surface is required immediately (in a road, for example) then a mechanical compactor can be hired but this can be avoided in most situations where patience and effort can be applied.

## 10.2   Forming openings in or taking down a wall

In the first place it is necessary to check the wall:

- What is the material and structure?
- What is its general condition (behind the plaster perhaps)?
- Is it load-bearing (carrying a floor, wall or roof above)?

Some materials such as cob, compacted earth or clunch have little structural strength and making new openings should be avoided. As we saw in Part 2.7.4, the redistribution of loads which occurs when large holes are formed in walls can overload foundations, but when dealing with weak materials there is the added danger that the materials in the wall will crack or even collapse. Much depends on the condition of the wall and this may not be revealed until the finishes have been removed. Coursed brickwork, blockwork or stone masonry has an inherent structure which can be supported while forming an opening. This is done using traditional methods involving the use of 'needles' and 'dead shores' as illustrated in Figure 10.2.

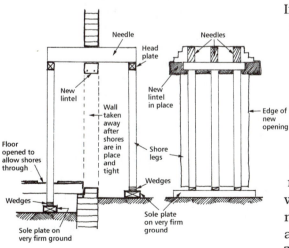

*Figure 10.2   Needles and dead shores*

If the condition of the wall is poor, though, and the opening large, a contractor may be the best person to deal with it. Collapsing masonry can not only cause structural damage to what it supports but damage a floor below and, much worse, injure those working on the job. Most new openings will require a new lintel to be inserted. These are available from builders' merchants in steel or concrete and in standard lengths. Cold-formed steel lintels of the Catnic type have the advantage of being light in weight but, whichever is used, the end bearing must be adequate. See Figure 10.3. Note the use of a 'padstone' which spreads the load from the beam into the wall below.

Timber framed walls may seem initially the most amenable to being opened up by the removal of infill panels. These may be brick or block, lath and plaster or, in very old buildings, wattle and daub. On a note of caution, the older the building the more likely it is that the panels are actually helping to support the structure, probably by stiffening an otherwise rather flimsy frame. If this seems likely the frame should be braced temporarily and the panels removed carefully until the full effect of the operation can be ascertained. Certainly if the formation of the opening requires the removal of frame timbers it will be necessary to 'trim' the opening with new timber and provide alternative bracing to the frame as illustrated in Figure 10.4. Where the timbers are crucial to structural stability, for example the struts of a roof truss, they should not be removed without the advice of a structural engineer.

Concrete walls present rather different problems. The material is very tough, particularly when reinforced with steel, so that cutting a hole in it can be done without the need for temporary support. The work should be done with a hand-held masonry saw but not without gloves, a safety helmet and ear and eye protectors. For a large opening a good number of cutting discs will be needed and the dust will be considerable. Concrete is very heavy, a 600 mm × 600 mm piece of 125 mm wall weighing over 100 kilograms, so the opening must be cut out piece by piece from the top. The pieces should be lowered to the floor rather than dropped, an operation best done by at least two strong people.

Load dispersed through padstone

Concrete padstone precast or cast in situ

* A bearing of 100 mm is typical. In special cases and where proprietary beams are used, the manufacturer's or engineer's advice should be taken

(a) Beam end bearing on padstone

(b) Lintel end bearing

Figure 10.3 Lintel or beam bearing

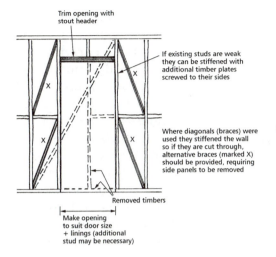

Trim opening with stout header

If existing studs are weak they can be stiffened with additional timber plates screwed to their sides

Where diagonals (braces) were used they stiffened the wall so if they are cut through, alternative braces (marked X) should be provided, requiring side panels to be removed

Removed timbers

Make opening to suit door size + linings (additional stud may be necessary)

*Figure 10.4    Forming an opening in a timber stud wall*

Removing a whole wall rather than part of it will bring similar problems to those mentioned above. It is, particularly important, though, to consider the following:

● The extent to which the wall is carrying a load from above. If it is load-bearing then a beam must replace it and the support for that beam must be carefully considered. If the wall is a cross-wall, the two walls at both ends may not be strong enough to bear the additional stress imposed by the beam. Moreover, one or both of them may be party walls requiring special consideration, as we saw in Part 5.1.2. Beams like lintels must have adequate bearing, which is why it is often necessary not to remove the whole wall but to leave piers in place or build new ones to absorb the additional stress induced under the beam ends. See Figure 10.5. Clearly the foot of a demolished wall up to floor level should be left in place so that this provides a base for the piers.

● The importance of lateral support. The wall to be removed will probably have a buttressing effect on the adjacent walls. One advantage of box-like buildings is that walls brace or stiffen each other. Otherwise a long wall will tend to buckle. This is another good reason for not demolishing the whole wall but leaving piers of a reasonable size to act as buttresses.

● Inserting a beam to replace a load-bearing wall requires the floors or roof above to be supported temporarily.

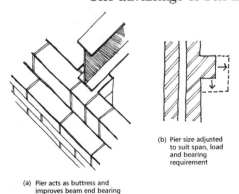

(b) Pier size adjusted to suit span, load and bearing requirement

(a) Pier acts as buttress and improves beam end bearing

*Figure 10.5    Buttressing effect of piers*

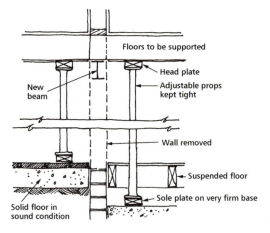

*Figure 10.6 Supporting floors before removing a wall*

This must be done with a robust system of props and plates carrying the load to solid ground. Using wedges and props of the Acrow type enables the whole thing to be kept tight so that when the wall is removed there is absolutely no downwards movement of the floor or roof. See Figure 10.6.

● The weight factor. Again it is necessary to consider the weight of material involved with demolishing a wall. A 100 mm concrete block wall, plastered both sides, 2.4 m high and 3.6 m long can weigh over 2 tonnes. Taking it down piece by piece, from the top, and lowering the blocks to the floor is the answer. Safe methods require patience and time but are essential in work of this kind.

## 10.3   Removing and restoring a fireplace

Removing a fireplace that is out of character is a common task. First get the chimney swept. It is better to hire an expert rather than the equipment – there is always a large quantity of soot which will fly everywhere. If there are no local sweeps, contact the National Association of Chimney Sweeps (NACS) at St Marys Chambers, 19 Station Road, Stone, Staffs ST15 8JP, tel 01785 811732.

Prepare for a lot of rubble. Unless the chimney stack has been removed do not remove the chimney breast in the rooms below without inserting some form of structural support. Even then, consult the local building control officer. The breast may be helping to stiffen a party wall and this will need special action, as we saw in Part 5.1.2.

Tiled fireplace surrounds are usually constructed in one piece, often with reinforcing bars embedded in concrete behind the tiles. The surrounds are generally fixed with lugs on either side screwed to plugs fixed in the wall. They are notoriously difficult to break up and, although very heavy, should be taken out in one piece.

The imprint of the surround and often of the original fireplace tend to remain visible even with careful patching. It is useful if the original outline remains as a guide to the proportions of any replacement. Once noted, it is best to be radical and hack off all loose or uneven plaster, clean off soot deposits and then use a proprietary sealer such as Unibond PVA to seal the surface before replastering. Using foil-backed plasterboard is an alternative and gets over the problem of the soot leaching through the new plaster, but plasterboard has a very crisp finish and may look so precise that it contrasts too much with the other finishes in the room. Paint manufacturers should be consulted where persistent soot staining occurs.

Architectural salvage yards usually have fireplaces but it may be difficult to find one of the right dimensions. If so, one made with old timber makes the best replacement. Medium density fibre (MDF) board can be a substitute as it is ideal for experimenting with traditional paint finishes. Fire resistant boards such as Masterboard are the ideal ground for a tiled surround inside the fireplace. The hearth must be built of incombustible material such as slate or tiles.

## 10.4  Plastering and plasterboard

Plastering is a difficult job for an amateur. The materials are heavy and it is necessary to work very quickly and in a precise sequence. An enthusiast could experiment with the aid of a good technical guide. Proprietary patching plasters tend to be expensive and difficult to use. Plasterboard is an easier material to achieve a good finish. It can be fixed to timber or proprietary metal battens or plaster dabs. The space between battens can be used for running electric conduit and insulation.

Foil-backed plasterboard, known by various proprietary names including 'vapourshield', has inherent insulation and vapour barrier protection. This can be beneficial where condensation is likely to occur. Standard plasterboard sheets are large and heavy, 2.4 m × 1.2 m (8 ft × 4 ft), but easy to cut with a sharp knife or saw. 12.5 m thick boards are required for half-hour fire protection round stairwells, 9.5 mm for other partitions.

Fixing should be done with proper galvanized plasterboard nails which have a tendency to spring out if the board is not held firmly in place. Screw-type nails counteract this. Fixing large sheets to a ceiling is ideally a two-person job.

The best technique is to use feather-edged boards, with scrim tape over the joints and then skimmed with a wide spreader over the joint. Most DIY shops stock what is necessary for the job. See Figure 10.7.

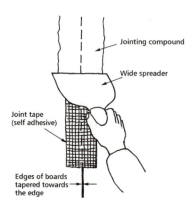

Plasterboard is the ideal material for internal partitions. It will carry considerable weights with toggle plasterboard fixings, has a high degree of sound insulation, is amenable to a variety of finishes and is relatively inexpensive. More insulation can be obtained by using 'thermal-check' plasterboard, i.e. 9.5 mm plasterboard bonded to 25 mm extruded polystyrene.

*Figure 10.7   Finishing a plasterboard joint*

### Plasterboard partitions

British Gypsum offer three partition systems which:

● are space saving
● are economical with timber
● offer good standards of sound insulation and fire resistance
● reduce to a minimum the amount of wet work
● are non-load-bearing.

They are:

● Gyproc Paramount dry partition. Lightweight partition 57 mm or 63 mm thick. Consists of two sheets of plasterboard bonded to a cellular fibre core. The sheets are 900 mm or 1200 mm wide × 9.5 or 12.5 mm thick. They are located by timber battens at the head and base and joined at battens pressed into the core at the side of each sheet. Services can be run in the core close to the battens, and battens can be inserted to take loads. The boards are tapered to allow panels to be finished flush with the Gyproc jointing system, or surfaced for Thistle multi-finish.

Paramount boards are not usually held in stock by builders' merchants, but can be ordered for quick delivery. This system is relatively easy for the amateur with the aid of British Gypsum's technical leaflets. See Figure 10.8a.

- Gyproc Gypwall. Lightweight partition 73 mm thick. Consists of two sheets of 15 mm plasterboard screw-fixed on either side of a light gauge metal framework. The framework has pre-cut holes for cables and pipes up to 25 mm. The installation of services and extra timber noggings for supporting shelves etc. is particularly easy with this system as only one side is fixed at a time. Finish as above. See Figure 10.8b.

- Gyproc laminated partition. Lightweight partition. Thickness varies from 50 mm to 95 mm according to the number of layers and thickness of boards used, which depends on the fire resistance and sound insulation standards required. It is constructed by setting the first board against 25 mm timber perimeter framing. Then a core of 19 mm Gyproc plank forms the centre of the sandwich stuck with bonds of Gyproc Dri-Wall Adhesive. Noggings to take fixings can be inserted in the core, also service pipes and conduit up to 25 mm thick. Plank or board is then fixed to the other side of the core with Dri-Wall adhesive. As soon as one layer is complete the next must be added, and when the wall is finished should be left to set for at least four hours and then for four days to develop full strength. British Gypsum state that erecting this type of partition is a one-person job. See Figure 10.8c.

**Note**

Full technical literature on the performance of these partitions and how to fix them is available from British Gypsum Limited,

(a) Gyproc Paramount partition

12·5 mm plasterboards give 63 mm partition.
9·5 mm plasterboards give 57 mm partition.
Can provide ½ hour fire resistance if given
2 mm thistle finish on both sides.
Sound reduction 30 decibels as shown.
Additional 12·5 mm boards on sides
gives 38 decibels (NHBC standard)

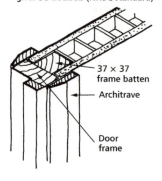

37 × 37
frame batten

Architrave

Door
frame

(b) Gyproc Gypwall partition

15 mm Gypwall board screwed to galvanized
mild steel channels, studs and noggings.
Can provide 38 dB sound reduction
and ½ hour fire resistance if properly built.

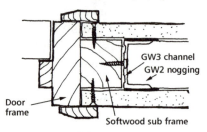

GW3 channel
GW2 nogging

Door
frame

Softwood sub frame

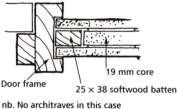

19 mm core

Door frame
25 × 38 softwood batten

nb. No architraves in this case

(c) Gyproc laminated partition

19 mm Gyproc Plank core with 12.5 mm
Gyproc Wallboard stuck on each side with
Dri-wall adhesive – gives 50 mm partition
capable of 1 hour fire resistance and sound
reduction of 33 dB.

Additional plasterboards can give 38 dB
as in (a) above.

*Figure 10.8 (a) Gyproc Paramount partition; (b) Gyproc Gypwall partition; (c) Gyproc laminated partition*

Technical Service Dept, East Leake, Loughborough, Leics LE12 6JT, tel: 0115 945 1000.

British Gypsum also offer short training courses at three centres in England: Erith, Loughborough and Penrith.

## 10.5  Building a partition

Partitions are usually walls which are non-load-bearing but must provide an effective screen against visual intrusion and sound. It is important that they are as imperforate as possible, so a degree of workmanship is required in their construction. Old houses often contain poorly built partitions using materials such as hardboard and fibreboard which are quite inadequate by modern standards. Fortunately they are relatively easy to remove.

Apart from the Gyproc type of partition described in the last section, there are two main methods of building new partitions: blockwork and timber frame. Both can be designed as load-bearing if required but their properties in other respects are rather different. As we have seen, block partitions are heavy and are best confined to situations where a solid base is available, i.e. usually at ground floor level. Brick or stone can be used where the 'fair-faced' quality of either is required but they require a somewhat higher level of skill and even a craftsman will find it difficult to achieve a fair face on both sides of a brick partition.

## Blockwork

Concrete block partitions have the advantage of providing good sound insulation provided the joints are completely filled with mortar. Blocks come in standard face dimensions of 440 mm × 215 mm which co-ordinate well with brickwork (two bricks long, three brick courses high) but there is considerable variety in thickness, from 75 mm up to 200 mm or more. Composition and prices vary too from the heavy aggregate blocks at about £6.50 for 10 to the lightweight aerated concrete type at about £8.50 for 10 (both prices relate to 100 mm thick blocks). They can be solid or hollow. The choice depends on the requirements, i.e.:

- Height and length of wall
- Thermal insulation (lightweight types are best)
- Sound insulation (dense types cope better with airborne sound)
- Strength (usually adequate)
- Fire resistance (usually adequate)

Appearance may be important where fair-faced work is required and some concrete blocks are made to resemble stone. The colour and shape of the joint then needs careful consideration – ideally the mortar should match the block in colour and be finished flush. Shrinkage of concrete block walls will occur, particularly if the blocks are 'green' and the wall gets very wet before being covered. This will result in cracking which may be harmless but can be alleviated by the use of a mortar no stronger than the blocks themselves so that the wall is to some extent flexible. Mortars containing a proportion of lime (1:1:6 or 1:2:9 cement lime sand) should be used in preference to pure cement mortars.

There are a number of DIY publications which advise on building block walls, including *How to lay walling blocks and bricks: Leaflet no. 16* from Homebase and *Constructing walling* from Do-It-All.

### Timber frame

A timber-framed partition consists of uprights (studs), horizontal members (noggings) and in some cases diagonals (braces). See Figure 10.9. The size of the timbers will be governed by the height and length of the wall. A strong wall capable of carrying loads will have 50 mm × 100 mm studs at 400 mm centres (spaced thus so that standard sheets of plasterboard 1200 mm wide can be fixed without cutting). Studs of 50 mm × 75 mm will be adequate for most domestic partitions provided they are suitably braced.

The stud partition is usually faced with plasterboard which should be 12.5 mm thick if sound insulation is required. Alternatively plaster laths (small plasterboards 300 mm ×

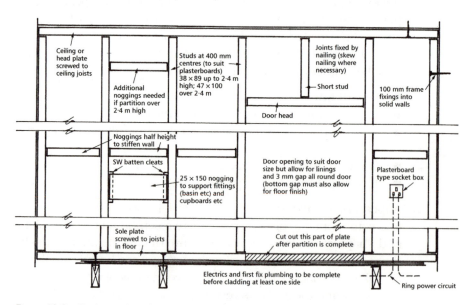

Figure 10.9   Timber and stud partitioning

1200 mm) or expanded metal lathing can be used as a backing for plaster finishes. Timber boarding is a popular finish but if fire resistance is required it should be fixed over a layer of plasterboard. Additional sound insulation can be provided by packing the gaps between the studs with a sound-deadening material such as mineral wool or fixing a double layer of plasterboard to one or both faces.

Advice on the building of stud partitions, including the level of skill required, the materials and equipment needed and so on, is available from the large DIY retailers. Typical are *Stud partition walls* from Do-It-All and *How to build a stud partition wall: Leaflet no. 18* from Homebase.

## 10.6   Wall tiling

Described by one DIY enthusiast as 'not particularly difficult but rather fiddly', wall tiling does need careful planning and an eye for detail.

It is a job which can easily be done in stages, and so lends itself to DIY.

Advisory leaflets from DIY retailers advise on the equipment needed. Do not economize on the tile cutter. A heavy duty cutter is essential for thick tiles but it also makes easier work of any tiles. See Figure 10.10. Tiles on a worktop should be set in water-proof adhesive and grouted with epoxy grout. This is expensive but more hygienic. Preferably use large tiles on worktops and small sizes on upstands. The larger the tiles, the less grout is needed, but large tiles are thicker and harder to cut.

If tiling for a shower compart-ment, first seal the wall with a sealant such as Unibond PVA to ensure adhesion and spread the tile adhesive over the whole surface with a serrated spreader. In less vulnerable locations, tiles can be fixed with dabs of adhesive close to each corner. This method is good if the background is uneven. Keep

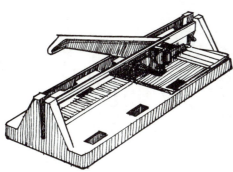

*Figure 10.10   A heavy duty tile cutter*

checking the surface with a rule to ensure it continues on an even plane.

Tile adhesives are quite potent, so work in a well-ventilated area. Work with gloves if possible and oil (baby oil) surplus adhesive off the hands at frequent intervals. Standard plasterboard is not a suitable background for tiling a shower enclosure. Use a moisture-resistant board or preferably exterior grade plywood which has been sealed.

Literature from DIY stores is available such as *How to fix ceramic wall tiles: Leaflet no. 02* from Homebase. The instructions on adhesive and tile packs are also very useful.

## 10.7 Electric wiring

Electric wiring is physically a relatively easy job for the amateur; however, the consequences of any errors could be dangerous and a good technical guide is essential. There is a wide range of materials and fittings readily available and DIY retailers can advise on the selection of materials and the tools needed. See Figure 10.11.

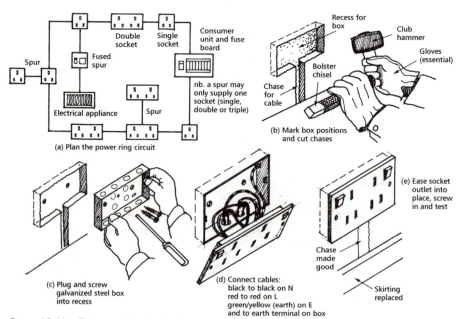

*Figure 10.11    Fixing an electrical socket*

It is particularly important to avoid running cables in places where it could be vulnerable to chafing or fixings. The cables should be run in conduit where there is a risk of them being damaged. In floors cables should be run through the centre of joists, not notched into the top. This not only prevents the weakening of the joists but prevents the cable from being damaged by nails or screws.

Horizontal runs in walls should be avoided; cables should run up and down to plugs or switches and fixings above them should be avoided. Hollow partitions such as those reviewed in Sections 10.4 and 10.5 make installation easier. However, it would be advisable to install the cable before at least one side of the plasterboard is fixed. As we saw in the previous chapter (Part 9.2.2), if other trades are involved the electrical work must be phased in with them: first fix work before linings, second fix before decorations and so on.

On completion, before wiring is connected to the main distribution board, the system should be thoroughly tested by a qualified electrician. The installation must be in accordance with the Regulations of the Institute of Electrical Engineers, whoever does it, and the test results must be satisfactory otherwise the supply authority will not provide a service.

Useful publications include:

- *The Which? Book of Wiring and Lighting*
- *How to upgrade a single power socket: Leaflet no. 10* from Homebase
- *Extra electric sockets* from B&Q

## 10.8 Plumbing installations

Any gas installation must be installed by a CORGI-registered installer. Ventilation is crucial and all instructions for gas fires and boilers must be observed. The consequences of not doing this could be fatal.

Plumbing for central heating and hot water is possible for the amateur with the aid of an efficient guide. Copper and plastic pipes and fittings are now used, the latter generally for waste runs.

*Joining copper pipes*. There are several types of fittings (see Figure 10.12):

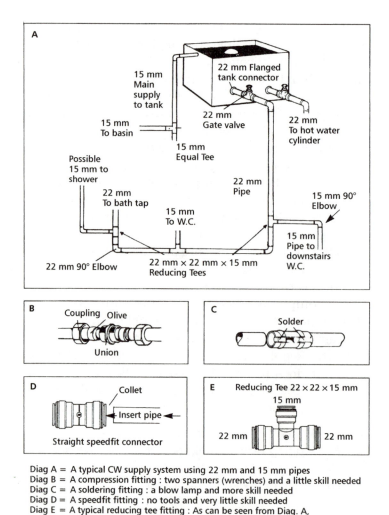

*Figure 10.12* *Plumbing fittings (based on the* Wickes Guide to Plumbing Skills)

- Comprehension fittings. Screw type – need no special tools but the joints are rather bulky and relatively expensive.
- Solder ring fittings. A solder ring in a coupling or fitting is heated using a gas flame to create a joint. These are much cheaper than compression fittings.
- Speedfit fittings. The fitting simply pushed into place. This is neat and simple although the fittings are rather expensive.
- End feed fittings. Fittings are cheap but practice is needed with the solder to form a watertight joint.

*Plastic pipe* is popular for wastes because it is easy to install, inexpensive, easy to wipe clean and does not discolour if short lengths are visible. Sometimes the threads on plastic pipe fittings fail if over-tightened, and discharging boiling water down wastes is best avoided.

As with electric wiring, care must be taken to ensure that pipe runs are not placed in vulnerable positions and that joists are not notched to impair their structural strength. Minibore pipework for central heating is a way of reducing these problems.

To avoid excessive deflection (sagging) or vibration, pipework must be supported on hangers or clips at regular centres. A horizontal 35 mm basin wastepipe, for example, should be supported every 1 m, a 15 mm copper waterpipe every 1.5 m.

Useful publications include:

- *The Which? Guide to Plumbing and Central Heating*
- *The Wickes Guide to Plumbing Skills: Leaflet no. 48* from Wickes
- B&Q, Homebase and Do-It-All also have leaflets on specific installations

## 10.9 Hanging doors

This is quite a complicated job for an amateur, requiring carpentry skills. It involves fitting hinges, bolts, latches or locks and handles, and may also involve:

- trimming the shape to fit the door swing (planing the edge)
- reducing the height to fit the opening
- fitting door closers.

Equipment required: a set of good carpentry tools.

To get a good fit the edge of a door is often bevelled to fit the door swing, so altering a door swing may involve retrimming a door, which reduces the stile or upright edge of the door each time. Fireproof doors are very heavy and should not be cut, because by doing so they may be no longer fire resistant. Ensure that the door opening is sufficient to take a standard full height door and normal floor finish to avoid making adjustments. Note that doors are still available in both metric and imperial sizes, with slight differences.

### Hinges

Heavy duty plastic covered steel with matching screw heads as well as steel and brass are available. Steel eventually rusts so brass is ideal for external locations. One and a half pairs are needed for heavy front doors – normally only one pair for internal doors.

### Locks

Plain mortice locks suit internal doors but for external doors a wide range of security locks is available. Home insurance companies will advise on what types they will accept. Fitting a new lock requires patience and some skill – the sequence is important. See Figure 10.13.

### Latches

Old doors are often too narrow for latch plates to be mortised into the door and surface fitted boxes are used instead. In a traditional setting these may look acceptable and even add character.

Useful references include:

- *How to Fit a Door: Leaflet no. 03* from Homebase
- *How to Fit Locks and other door furniture: Leaflet no. 04* also from Homebase

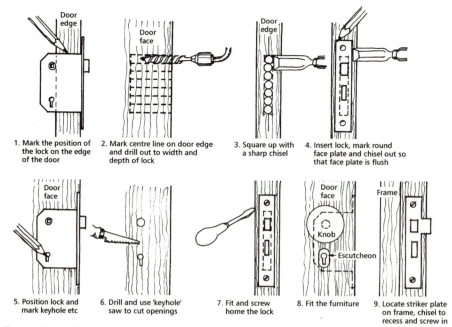

1. Mark the position of the lock on the edge of the door

2. Mark centre line on door edge and drill out to width and depth of lock

3. Square up with a sharp chisel

4. Insert lock, mark round face plate and chisel out so that face plate is flush

5. Position lock and mark keyhole etc

6. Drill and use 'keyhole' saw to cut openings

7. Fit and screw home the lock

8. Fit the furniture

9. Locate striker plate on frame, chisel to recess and screw in

*Figure 10.13   Fitting a door lock*

- *New Door Locks* from B&Q
- *Fitting a front door* also from B&Q

## 10.10  Shelving

Keep any old shelves because they may have more character than new ones. Strip paint and reposition them as required. Stripping paint is a noxious and messy process and slow. There are eco-friendly stripping solutions but it is preferable to have work done commercially.

There are innumerable proprietary shelving systems, some with visible supports such as aluminium brackets, some with discreet fixings such as 'tonks' fittings, and some with wholly invisible supports produced in ready-made form by the DIY chains. Mounting instructions are generally included, or fixing diagrams displayed. Ready-made shelves come in a wide range of sizes but will almost always have to be cut to fit. See Figure 10.14.

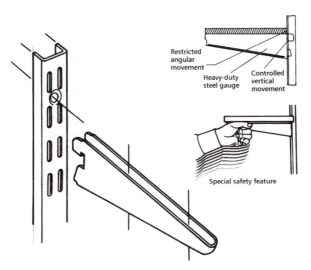

Restricted angular movement

Heavy-duty steel gauge

Controlled vertical movement

Special safety feature

The Spur 'Steel-Lok' adjustable shelving system – a DIY favourite for many years. Double-slotted steel uprights support steel cantilever brackets of different sizes all finished in powder coated epoxy/polyester paint in a range of colours. The system can be used for a wide range of shelf types. Properly built it will carry loads of up to 55 kg per bracket.

*Figure 10.14    The Spur shelving support system*

### Making shelves

Ensure the thickness of the board and spacing of supports is sufficient to take the weight of the items stored without deflecting. Standard plastic-coated shelves usually sag under the weight of books so supports as close as 600 mm may be necessary. Both timber and MDF board are suitable, dependent on the character required. Pine-board, made of laminated strips of timber, was devised to reduce warping and so is very suitable. MDF board lends itself to simple screw and glue techniques, is very strong and does not warp. It is also an ideal background for decorative techniques.

Note: Masks should be worn for sawing MDF board – and it tends to blunt tools. Some DIY stores will cut MDF to size. A charge is made for each cut but the precision and transportability are a great advantage.

Useful leaflets include:

● *How to Use Pineboard: Leaflet no. 38* from Homebase. This leaflet includes instructions on constructing a small bookcase, listing tools and materials appropriate for constructing shelving generally.

## 10.11  Painting and decorating

These activities have been the most popular DIY jobs for a very long time, so there are many publications and courses available. Even beginners can make a start by experimenting on small areas before tackling a whole room. The following notes are intended as a guide only to some of the practices involved.

### Materials

To appreciate the wide range of decorating materials available, spend some time browsing through the shelves of a DIY shop. Emulsion and resin-based paints are now common and of reliable quality. There is an increasing trend towards using water-based paints and glazes because they produce no noxious fumes or solvents. Water-based paints dry quickly which is usually an advantage but does not suit some decorative techniques. The ability to wash out brushes and splash marks with water is a considerable advantage but it has be done quickly. Once dry, emulsion paint is difficult to remove from most surfaces. Try baby oil to remove it from the hands. Where oil- or resin-based paints are used, clean surplus paint off brushes thoroughly with newspaper and then a proprietary brush cleaner, not white spirit, so that brushes can be rinsed in water.

### Preparation

Preparation of the surfaces to be decorated will often take more time than the decoration itself. The work should not be skimped. Having the right equipment is a start: a stout pair of steps, preferably aluminium for lightness in weight and having a top platform and handrail, or for larger jobs a pair of trestles and at least three scaffold boards which can be hired from a plant hire company. For outside work the correct use of a ladder is essential, as we saw in Figure 2.3. If no one is available to foot the ladder, the bottom must be anchored in such a way as to prevent it sliding. Pegs, stout timbers and rope may be needed to do this. See Figure 10.15.

Stripping old paint or wallpaper is tiresome work which can be alleviated to some extent by the use of hot air blowers, steam strippers or chemicals. All these are potentially dangerous and must be used with great care. If a steam stripper is hired or bought, it is essential that the manufacturer's or supplier's instructions are followed to the letter. Remove obstacles from the work area; cover vulnerable surfaces and have a stack of old newspapers handy. Large dustsheets are useful. Wear protective clothing. Careful measuring of the surfaces to be decorated is advised. If colours or patterns have to be matched, take a sample to the shop – the lighting there will be different from that on site. If colour is to be mixed, make sure that

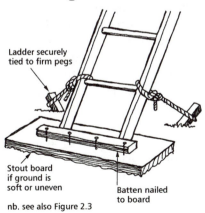

Ladder securely
tied to firm pegs

Stout board
if ground is
soft or uneven

Batten nailed
to board

nb. see also Figure 2.3

*Figure 10.15    Stabilizing the base of a ladder*

enough is prepared in one batch – running out and having to match a second batch can be difficult. In wallpapers, check batch numbers and when measuring allow for joints to occur at internal corners, particularly if the batch number has changed.

There is often a strong temptation to overpaint old surfaces, and in the case of walls and ceilings this is often satisfactory provided a light colour is not being applied to a dark background. However, with joinery, particularly window frames, door architraves and skirtings, too much over-painting destroys the crispness of detail which is so attractive. Where this has already happened, it is a better job if the old paint layers are removed using strippers or hot air (beware in the latter case of the need to shield glass from hot air). Sugar soap is a very good proprietary material for cleaning old paint surfaces where it is decided not to strip them off.

New joinery should be given the traditional treatment of knotting (applying shellac to knots), stopping (filling holes and cracks with filler) and priming (an essential base for all opaque finishes). If the finish is to be translucent paint, the material is applied straight to the unprimed surface.

### Working

Applying paint is usually by brush or roller but spray equipment can be hired. Following the manufacturer's instructions usually gives a satisfactory job. Where several coats are involved, ensure that they are all part of the same manufacturer's system. Pouring a usable quantity of paint into a paint kettle and keeping the lid on the main tin is a good idea. It prevents the paint from drying out or getting dirty and the loss of paint when the tin gets knocked over. There is a temptation to reduce the number of coats recommended by the manufacturer. This should be avoided. One-coat paints are available but others benefit from having a good undercoat because it is the thickness of paint which in most cases leads to long life.

Wet paint brushes or rollers can be kept in plastic bags secured with an elastic band for short periods. If they are neglected and become hard, discard them rather than go through an expensive and noxious process of cleaning. Consider using inexpensive throw-away brushes but discard those if they do not give the desired finish.

All the DIY chains produce leaflets on interior and exterior painting techniques, paperhanging and so on. Which?, Collins and Readers' Digest manuals cover the topic very well and for special effects books by Jocasta Innes and Kevin McCloud add inspiration. See:

- Innes, J. (1986) *Paintability*. Weidenfeld and Nicholson.
- McCloud, K. (1996) *Decorating Book*. Dorling Kindersley.

## 10.12 Clearing up and hiring skips

Moving rubbish can be heavy physical work. Take care and employ labour if not used to it, or a chiropractor will be the most likely to benefit. Otherwise masks, industrial gloves, stout boots, barrier cream and overalls are needed. A wheelbarrow is still a most efficient way of moving rubbish and can cost as little as £25 to buy.

### Removing loose plaster

The first three bucketfuls will almost fall off. After that it is usually hard going and very dusty. A wide chisel (bolster) and a club hammer are the basic tools – or hire electric tools. Lock and seal doors to rooms which are not to be disturbed with masking tape and dustsheets. Clear up systematically to prevent hazards accumulating.

### Transporting rubbish

Arrange for a skip. These can usually be hired locally at a cost of about £100 for a 6 yd (4.6 cubic metres) size and £60 odd for a 'mini-skip'. Large skips sometimes have a 'drop end' which makes them easier to load. If they have to stand in the road, notice has to be given to the local authority and a licence obtained. Suppliers of skips normally arrange this but they need several days' notice. Keep a space clear in the road ready for the skip. Hire it for as short a time as possible or it will be used by the neighbourhood for dumping rubbish. Clear up systematically; it makes a job run smoothly. If small quantities have to be transported by car, ensure that anything liquid is in buckets with tight lids and line the vehicle with plastic sheets. Hiring a van for a day is often an attractive alternative.

# Case studies

## 11.1   Case study 1 – Aldborough Road

This study illustrates the process of buying a run-down terraced house, improving and extending it, and finally selling it in a falling market. The venture only broke even financially so could not be considered a success for the developer but it left a series of pointers for a more profitable scheme which followed. In addition, the project provided employment for people at a time when the building industry was depressed, raised various taxes for the government and resulted in a more attractive and comfortable family house.

### Appraisal

A preliminary appraisal suggested that the house, at an asking price of £63 000, could be a viable proposition. At the peak of the property boom, similar houses in the locality, modernized and extended, were selling for around £155 000.

The location was good, convenient for town centre shopping, leisure and transport facilities, and in a quiet cul-de-sac close to the River Thames with a potentially pleasant garden and the bonus of a fine apple tree. As a residential area it had recently undergone a transformation and was still improving. See Figure 11.1.

In early 1991 the Halifax had forecast a rise of 5 per cent in house prices for the year. The estate agent too was quite optimistic about the property market, reporting one or two recent sales, including one for a similar house nearby at £135 000.

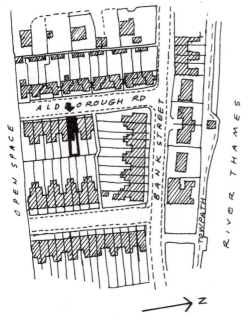

Figure 11.1    The location plan – Aldborough Road

They expected to achieve the same price again, particularly as there was a shortage of property for sale.

### Survey

A survey soon revealed the reasons for the apparently low price. Building work had been started on the house and then abandoned, leaving the rear extension without a roof. The house was a Victorian terraced house but the original fireplaces and balusters had been removed and much of the plasterwork was in poor repair. There was no dividing wall in the attic to separate adjoining houses.

However, some positive work had been done. Drainage for the kitchen and bathroom had been started on the assumption that they would be in the rear extension. The main roof had been overhauled and insulated. Access to the attic had been created by reversing the staircase and the first floor ceiling timbers had been replaced by floor joists at a lower level, to take heavier loads and to improve headroom in the attic.

The timber floors were generally in good condition and the windows capable of repair. Most of the original doors were acceptable but the ironmongery was in poor condition.

New heating, plumbing and electrical installations were required and plans for a new kitchen and bathroom needed, but firstly a decision had to be made on whether to build an extension into the garden because this would affect the location of the kitchen.

Detailed dimensions were taken and measured survey drawings prepared. See Figure 11.2.

### Outline proposal

The proposal was to buy the house for £63 000 and upgrade and extend it for resale. The total cost of the works and the

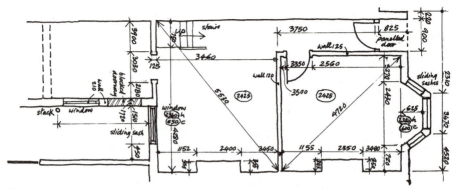

Figure 11.2    The ground floor survey drawing

house, including buying, financing and selling costs, was not to exceed £100 000. An estimated selling price of £135 000 seemed to allow a comfortable margin for unforeseen extras and a good profit.

A scheme was suggested by the fact that the rear of the house faced east and the garden was potentially a feature of the property. The orientation caused overshadowing close to the house so it was decided to construct a single-storey extension as a family room which would allow people to emerge into a sunny garden in the late afternoon. See Figure 11.3.

It was considered important to locate bathrooms with easy access to bedrooms, so one was planned next to the attic bedroom. See Figure 11.4. This suggested that another bathroom on the first floor could be quite small. Steps on the landing meant that the safest location for the majority of the occupants would be between the two main bedrooms. See Figure 11.5.

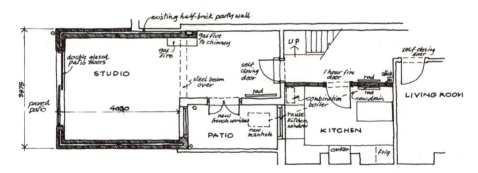

Figure 11.3    The ground floor proposals

Finally it was decided to combine the kitchen with a breakfast room, so the kitchen was moved from its old position at the rear to the larger room in the centre of the house. A patio for growing herbs alongside a more formal dining area separated the kitchen from the proposed extension. See Figure 11.3.

It was anticipated that the works would take three months and that the house would be ready for resale at the beginning of September. A key factor was the decision to ensure that the extension was within the volume of permitted development, thus saving time.

Plans and approximate costing for this scheme were prepared and a cash-flow projection to show what finance was required and the stages it would be needed. See Table 11.1.

### Obtaining finance

The developer first approached his bank with the scheme and cash-flow projection. Relations with the bank were quite good because previous ventures had been successful. However, the

**Table 11.1    The cash-flow projection – Aldborough Road**

*Aldborough Road – Cash-flow projection Feb. 1991*

| | 1991 March | April | May | June | July | August | September |
|---|---|---|---|---|---|---|---|
| *Expenditure* | | | | | | | |
| Purchase of property | 6200 | 55800 | | | | | |
| Legal costs, stamp duty, etc. | | 1500 | | | | | |
| Construction costs | | | 10000 | 10000 | 2500 | | |
| Interest on loan | | 100 | 900 | 1000 | 1100 | 1150 | 1150 |
| Cost of selling (fees) etc. | | | | | | | 3000 |
| Monthly total | 6200 | 57400 | 10900 | 11000 | 3600 | 1150 | 4150 |
| Cumulative total | 6200 | 63600 | 74500 | 85500 | 89100 | 90250 | 94400 |

| *Income* | |
|---|---|
| Sale of property | worst case          110000 |
| | best case            130000 |
| Therefore profit (before depreciation) | worst case  110000 – 94400 |
| | =      15600 |
| | best case    130000 – 94400 |
| | =      35600 |

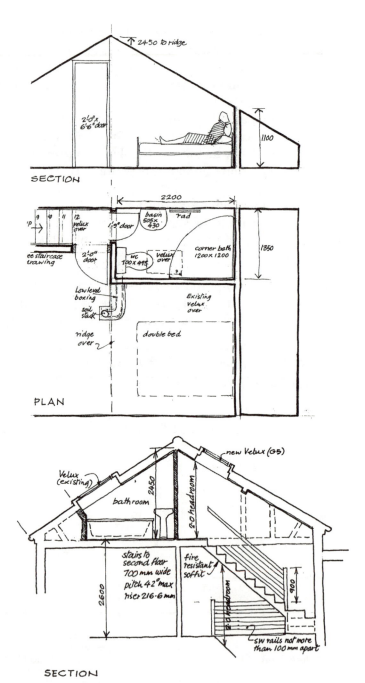

SECTION

PLAN

SECTION

Figure 11.4    The attic conversion

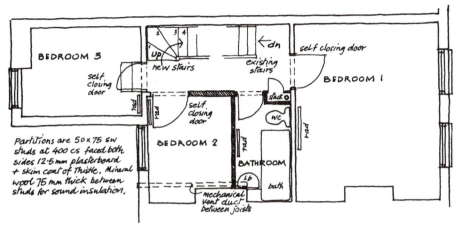

*Figure 11.5    The first floor bathroom*

property market had experienced a severe downturn and the new terms reflected this. A loan of £92 000 was requested. It was offered on the following terms:

- An arrangement fee of £900
- Rate of interest to be 3 per cent above Bank base rate (i.e. 13 per cent or 16 per cent in total)
- Security to be provided by legal charges on three properties

A first charge was required on the house to be purchased and on a holiday home valued at £45 000 owned by the developer, and a second charge (because it was already subject to a mortgage with a building society) on the developer's home. An estate agent's valuation would have been acceptable for the first two properties but a professional valuation was required for the latter. This would have incurred valuation and legal fees, both subject to VAT. Assuming a period of six months for the loan and a conservative figure of £1500 for the combined fees and VAT, the cost of finance for the anticipated period would have been £9000.

This was considered disproportionate to the project as a whole so the developer decided to approach his building society.

In March 1991 the building society was prepared to offer bridging finance for the project at a rate of 12.9 per cent as an extension of the existing mortgage on the developer's house. This rate was relatively low because it was subject to a large loan discount of 1 per cent. A valuation survey had recently been done on this property so no additional fees were requested.

This reduced the projected cost of finance to approximately £6000 for six months. This still seemed a high proportion but it was thought manageable and the scheme proceeded on this basis.

Before accepting an offer for the property, the estate agent asked for an assurance that the purchase would be for cash because the state of the property could have made it unmortgageable. In the event it might have been difficult to meet the terms of the bank's offer because another agent might have been reluctant to give a valuation.

### Problems between exchange and completion

Party wall procedures were not regulated outside London before 1996 so no survey had been conducted in relation to the party wall before the previous building works began. The fact that a joist had been inserted into a wall only a single brick thick did not come to light until the period between exchange and completion. The builder had disappeared, leaving the old lady next door to worry about the damage. See Figure 11.3. In fact the damage was slight and was rectified once the new work started.

A greater problem was the fact that in the period between exchange and completion heavy rain had affected the rear wall of the main house on the side of the unroofed extension. The building control officer insisted that part of the wall was to be underpinned and a new foundation constructed. The additional cost was estimated at £1700 and the work extended the programme.

It should have been possible to negotiate a reduction in the purchase price since the vendor was still responsible for the state of the property, but the vendor was obstinate and the need to act fast to prevent any further deterioration of the fabric left no time to press the matter.

### The building works

Plans and the required fee were submitted to the Building Control Department. They asked for extra information on several points and then stamped the plans as approved.

The developer decided to negotiate a contract with a builder who had been highly recommended by an acquaintance and whose work was to a good standard. He was interested in doing the job and was able to fit it into his programme.

Copies of the approved plans and a schedule of works were sent to the builder together with a timetable of operations. The schedule was priced and agreement was reached on how extras were to be calculated and on subtracting work from the contract if necessary in order to keep within the budget for the job. See Tables 11.2 and 11.3.

Work was started and proceeded broadly according to plan. As usual in such jobs, some unforeseen extras were necessary and minor modifications were required to add visual interest, such as French doors to the patio. These were priced as agreed and balanced by savings in taking some items from the works.

Towards the end of the contract, work slowed down and then was caught in the annual holiday period. Decorating and part furnishing followed. Eventually the house was ready to put on the market in October, a month later than had been planned.

### Selling the house

Vendors had been encouraged by activity early in the year and over the summer quite a number of properties had been put on the market. A slowdown was now evident and the agents were slightly less optimistic. Nevertheless it was agreed to put the house on the market at the price originally discussed with the same agents. The agent's fee was to be 2.25 per cent for selling the house on a sole agency basis for a period of 10 weeks.

The house was put on the market for £135 000 towards the end of October 1991. See Figures 11.6 and 11.7. Potential buyers offered £125 000 almost immediately. It was too close to Christmas to prevaricate and with the offer of an early exchange of contracts would have been acceptable, but the prospective purchasers were unable to sell their own house.

Newspapers reported that the lift in property prices and transactions evident in the spring had fizzled out and then reversed into a slide in the autumn of 1991. The days turned colder and darker and the extension, which had been designed to make the most of the garden, appeared cold in an unoccupied house over the winter. The estate agent recommended making the house look cosier with warm-coloured rugs.

An attempt was made to let the house and the agents were very optimistic, but letting, like selling, has difficult periods and there was no reason to suppose that the market for selling would

## Table 11.2    The part schedule of work given to the contractor

*Procedure*

The contractor is to complete the construction of the walls of the rear bedroom (first floor), replace the roof to a watertight state, and provide at least temporary rain water disposal BEFORE proceeding with any other work.

*Decorations*

The contractor is asked to provide separate prices in his tender for decorations, i.e. painting and ceramic tiling.

*Demolition, excavation etc.*

1. Demolish short length of brick wall in the back area and make good exposed end of wall to be retained.
2. Excavate for drainage along lines shown on the drawings. Backfill and consolidate after laying drains. Take care not to damage the existing drains or other services.
3. Remove and cart away the timber floor joists, boarding, plates, bearers, etc. comprising the timber floor in the back room, ground floor.
4. Cart away surplus rubble existing on site (but see clauses 5, 10 and 18 below in which some of the rubble will be reused).
5. Excavate for foundation to new back extension. The bearing capacity of the formation should be not less than 25 kN/m. Provide and lay a bed of compacted hardcore as necessary.
6. Remove existing soil and vent pipe and waste pipes in the back yard and cart away.

*Area under existing timber floor*

7. Take note that the height from the ground surface to the top surface of the existing boards is approximately 1350 mm.

8. Remove by pumping or other means the water lying on the ground surface.
9. Remove all rubbish and organic material from the area, down to a firm surface.
10. Lay and compact over the area approximately 150 mm of hardcore. Brick from the existing pile of rubble may be used provided it is well broken and free from non-brick material. Blind with dry lean mix.
11. Cover area with Visqueen 1000 sheeting turning the material up at least 150 mm at edges.
12. Lay 100 mm concrete to a level top surface over the whole area and trowel smooth.

*Drainage*

13. Lay 100 mm soil drains in UPVC pipes and fittings, including new inspection chamber in patio.
14. Connect new drainage to existing system in a manner to be approved by the local Building Control Officer.
15. Depth and falls of drains to be determined by position and depth of existing drains but falls generally should not be less than 1:60.
16. Provide suitable connectors for wastes in sink and washing machine locations and a rodding eye at the point of connection with the Soil and Vent Pipe.
17. Lay 100 mm UPVC surface water drains from the position of the RWP in the patio connecting to the existing SW drain and from the position of the new RWP on the new rear wall to a soakaway.
18. Form soakaway, to consist of 1½ cubic metres of clean brick rubble topped with polythene sheeting and at least 300 mm of top soil.

**Table 11.3  The part schedule of work priced by the contractor**

| Job no. | Schedule Works | Value £ |
|---|---|---|
| prlms | Complete construction of rear extension roof | 693.00 |
| 1 | Demolish short length of brick wall | 50.00 |
| 2 | Excavate for drainage | 130.00 |
| 3 | Remove and cart away timber floor joists | 60.00 |
| 4 | Cart away surplus rubble | 120.00 |
| 5 | Excavate for foundation | 300.00 |
| 6 | Remove soil and vent pipe | 40.00 |
| 10 | Hardcore and blinding to under floor | 240.00 |
| 11 | Damp proof membrane to under floor | 40.00 |
| 12 | Concrete to under floor | 300.00 |
| 13 | Lay soil drains and construct new man hole | 574.00 |
| 17 | Lay surface water drains | 55.00 |
| 18 | Form soakaway | 140.00 |
| 19 | Concrete to footings | 512.00 |
| 20 | Concrete to oversite | 150.00 |
| 21 | Concrete to back room | 80.00 |
| 23 | Cavity walls (bricks priced @ £300 per 1000) | 1,867.00 |
| 25 | IG Ll/E WIL lintel to patio doors | 58.00 |
| 27 | Drylining to neighbouring kitchen | 215.00 |
| 28 | Roof to extension incl. insulation, glazing etc. | 1,463.00 |
| 30 | Padstones to purlin | 20.00 |
| 31 | Tiles and screed floor to extension | 648.00 |
| 32 | Supply and fix Elbee patio door | 550.00 |
| 33 | Chipboard to kitchen floor £120, PC repair £100 | 220.00 |
| 34 | Construct stud partition | 270.00 |
| 35 | Supply and fit 1 hour fire resisting door | 140.00 |
| | Total this page: | 8,935.00 |
| | Subtotal: | 8,935.00 |
| | VAT @ 17.5%: | |
| | Estimated total: | |

be any better at the end of an average lease, and it might even have been worse, so the idea was abandoned. The projected rent of £775 per month would not have covered the interest on the loan.

An opportunistically low offer was made but was not accepted. Eventually several more realistic and acceptable offers were received, but not until the following spring. In May 1992 the house was finally sold for £115 000. In the meantime interest on the loan had mounted steadily and the cost of finance was second only to the cost of the building work.

### Evaluation of the project

Extra costs were incurred in several areas but the most significant ones were due to the delays and pressures caused by the

| | |
|---|---|
| HALL: | Tiled flooring. Understairs plumbing for washing machine. Radiator. Power points. |
| FRONT RECEPTION: | c. 13'4"x10'6". Open fireplace. Stripped pine cupboards to recesses. Bay sash window to front. T.V. aerial point. Radiator. Power points. |
| SITTING ROOM/ STUDIO: | c. 21'10"x11'0". Double glazed patio doors to garden. Ceramic tiled flooring. French doors to to courtyard gravelled area to side. Sash window to courtyard and South facing light giving window to side. Gas fired combination boiler for central central heating and hot water. Radiator. Power points. A delightfully bright room. |
| KITCHEN: | c.11'3"x8'0". Well fitted with attractive range of white wall and base units providing ample storage and work space. Inset sink and drainer. Window above overlooking courtyard. Fitted electric oven and gas hob. Feature arch and extractor over. Concealed lighting and spotlights. Plumbing for dishwasher. Radiator. Power points. |
| FIRST FLOOR: | |
| LANDING: | Stairs leading to second floor. |
| BEDROOM 1: | c. 13'9"x11'4". Twin sash windows to front. Radiator. Power points. |
| BEDROOM 3: | c. 8'11"x6'11". Sash window to rear. Fireplace recess. Radiator. Power points. |
| STUDY/ BEDROOM 4: | c. 8'2"x6'5". Sash window to rear. Fireplace recess. Radiator. Power points. |
| BATHROOM: | White suite panelled bath, low level w.c. and pedestal wash hand basin. Extractor fan and light. |
| SECOND FLOOR: | |
| BEDROOM 2: | c. 15'4"x9'10". Velux windows to front and rear. Eaves storage cupboard. Radiator. Power points. |
| BATHROOM: | White suite with corner bath, low level w.c. and pedestal wash hand basin. Extractor fan and light. Radiator. |
| OUTSIDE: FRONT GARDEN: | Enclosed with brick built wall to front, tiled path to front door and garden to side with attractive plants and shrubs. |
| REAR GARDEN: | Paved patio. Area of grass, trees and shrubs. Enclosed with fencing to sides and wall to rear. |

G E N E R A L L Y:
All Mains Services are Connected. Vacant Possession on Completion.
PRICE..........................£135,000....................FREEHOLD
(SUBJECT TO CONTRACT)

Figure 11.6    The sale particulars

*Figure 11.7    Views of the completed work*

depressed state of the housing market; this lowered the selling price and extended the period for which the loan was required.

Other problems were that the house required rather more work than had been apparent in the preliminary survey, in particular to the foundation of the rear wall of the main house. The extra work added to the original programme may have been a factor in the property eventually reaching the market too close to the end of the year.

Lastly the extension may have been too ambitious for the general character of the house. It looked attractive and had added to the interest of the project but it probably did not add its cost to the immediate value of the house. A more conventional solution keeping the kitchen in the rear extension would have cost less, been marketed at a lower price and hence would be likely to have appealed to more people. It would also have been finished more quickly and would have arrived on the market at a more favourable time of year.

It may be that people are more conservative in their tastes when the housing market is depressed; they certainly feel they have less money to spend.

However, from this experience a series of pointers were established for the next project.

### Pointers for the next project

- Take account of interest rates. It is difficult to achieve a successful outcome for any project with interest rates at over 10 per cent. Wait for a more helpful financial climate.
- Choose a time when the market is on the turn and poised to go up. This is difficult to evaluate but an instinctive feeling may be more useful than statistical projections. A falling property market will almost inevitably cause delays, encourage buyers to squeeze the price and undermine confidence in the project.
- If a scheme is entirely for development and not for personal use, the amount of work should be kept to a minimum in order to limit expense, time and interest on any loan. Other people are bound to have their own ideas of the ideal extension; the main point is to provide an attractive base which they will want and from which they can start.
- If unforeseen factors depress the market and upset the timing, then make a decision to let the property as soon as possible, planning for the end of the tenancy to occur early in the year if at all possible. This could have helpful tax implications as well.
- Negotiate lower fees with estate agents as a developer.
- Consider more DIY. This can mean anything from conveyancing, managing the job or doing some trades.
- It is important to maintain momentum. It is realistic to assume that if the employer takes a short holiday then everyone else involved with the job is likely to take a long one.

## 11.2  Case study 2 – Hook Street

This study starts from a different basis. The prospective buyers here were looking for a home for their retirement close to the town centre and with easy access to riverside walks. See Figure 11.8. Eventually it became a development project for two reasons: retirement was delayed and interim letting had become a problem.

### Appraisal

The two-storey terraced house, built about 1885, was on offer at £84 950 in the early summer of 1995. Property prices were

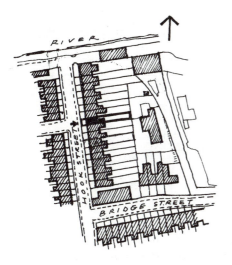

Figure 11.8    The location plan – Hook Street

still drifting downward (and continued to do so until about August that year). The property market felt very depressed and activity was low.

From its external appearance the house appeared to have lost some of its original character and the houses on each side had a few shabby features. Industrial premises and an empty building at the end of the street added to a gloomy prospect. However, the location was very good, the neighbouring houses seemed to have repairs pending and the purchasers felt convinced that the market was about to bottom out.

## Survey

A measured and condition survey was undertaken. See Figure 11.9. Externally the roof and walls, gutters and downpipes were in good repair. The original sash windows had been replaced with aluminium ones with a fixed lower light and a top opening one, and were only single glazed. A door of the 1930s era replaced the original one. The interior was permeated with strong cooking smells from a dark and greasy kitchen and the colours of the carpets and wallpapers generally were rather overpowering.

On the ground floor the original doors had been replaced with flush faced firecheck doors, possibly in preparation for a loft extension which had never been built. The original doors were still on the first floor, with tack marks all round the edges where hardboard had been pinned. Fortunately the original balusters, albeit pinmarked, also remained, but none of the original fireplaces.

A bathroom had been installed on the ground floor beyond the kitchen, and hot water was supplied from a 'fortic' tank in the bedroom above. A timber-framed plastic-roofed enclosure in poor repair protected the back door and a clothes drying area.

Heating on the ground floor was by electric storage units and on the first floor by electric panel heaters. The electrical

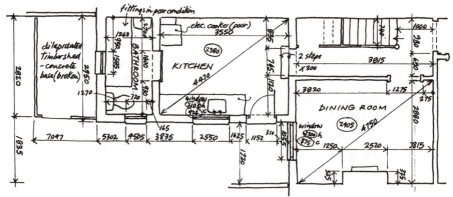

*Figure 11.9   The ground floor survey drawing*

installation seemed to have been recently overhauled and was rather sparse, but sound. There was no mains gas supply.

The suspended timber ground floor was well ventilated and in good condition, as were the upper floors.

### Cost plan

Discussions were held with estate agents in relation to letting the property after improvements.

Estimates for rental varied from £700 to £900 pcm so £800 was assumed. On this basis it was decided that the total cost of the project should not exceed £80 000. Allowing for agents' fees and vacant periods, this would be a yield of 10 per cent per annum and therefore cover the mortgage and produce a small profit.

Although the asking price for the house was virtually £85 000, its appearance and condition, coupled with the state of the market, suggested a much lower figure. So a considerably lower offer of £73 000 was assumed, leaving a total of £7000 for the works. Building work was to be finished quickly and the house let by 1 September.

Plans were drawn up with the intention of doing the work in two stages. Stage 1 was for immediate letting, Stage 2 was for later occupation. Work for the first phase had to take account of the eventual plan. A schedule for the first stage was prepared and approximately priced. See Table 11.4.

To keep costs to a minimum, as much work as possible was to be DIY, including conveyancing. The scheme was considered

### Table 11.4   The priced schedule of works – Hook Street

*Schedule of works and approximate costing*

| Item | Cost (£) |
| --- | ---: |
| Gas supply (living room, dining room, kitchen) | 150 |
| Kitchen units, cooker and fridge | 1000 |
| Conservatory | 2000 |
| Bathroom | 500 |
| Cloakroom suite | 150 |
| Kitchen flooring 10 sq yds | 150 |
| Inspection chamber, soil stack and drains | 600 |
| Plumbing | 600 |
| Concrete base for conservatory | 200 |
| Cloakroom partition | 50 |
| Assemble kitchen | 300 |
| Assemble conservatory | 300 |
| Partition for bathroom | 200 |
| Opening for bedroom door | 200 |
| Doors for bathroom and WC | 100 |
| Refit kitchen door | 20 |
| Conservatory 2 doors | 200 |
| Electric light and vents for baths and cloaks | 300 |
| Renew cooker control panel | 50 |
| Extra sockets kitchen utility and conservatory | 250 |
| Light to conservatory | 50 |
| Total | 7370 |

Note: The conservatory was later purchased on a no-interest loan over two years, so this item was subtracted from the initial costs and economies were made on several items, including the kitchen units which were bought as a special offer.

too small to require a general contractor so direct labour was employed for the specialist trades. As a fall-back position, estate agents were asked for a price range for selling after modernization and a range of £115 000 to £130 000 was suggested.

### Design scheme

The work was divided into two stages, a minimum amount to be done immediately, followed by an attic conversion when the owners were ready to occupy the property. The attic was eventually to become a large bedroom with an en suite bathroom, similar to that used in Case study 1 (Figure 11.4).

The kitchen was gloomy because it looked out onto a narrow yard and was blocked from the garden by the bathroom, so it was decided to locate a new bathroom upstairs and turn the old one into a utility room linked to the garden with a conservatory. See Figure 11.10. Conservatories do not require approval under the Building Regulations and the small conservatory was within the parameters of permitted development in the planning context.

Installing a small bathroom in the centre of the first floor meant reorganizing the hot water supply, so it was decided to install a gas-fired combination boiler. The central bathroom could be ventilated with a duct through the roof space above. See Figure 11.11. A cloakroom was installed at the rear of the kitchen, opposite the back door, as shown in Figure 11.10.

Leaving the kitchen open to the utility room and fitting a glazed door into the conservatory was to give a bright feeling to the kitchen. The design was 'galley style' with white worktops and an inset sink under the window, with white glass-fronted storage units on the opposite wall and highly reflective cupboards below. Spotlights over the worktop completed a crisp effect.

### Obtaining finance

Because the project was planned for their eventual home, the developers decided to approach their building society for a loan of £80 000. Although they had the equity to secure the loan they did not have the income to cover the extra monthly repayments unless the income from the property was taken into account. The building society was not prepared to do this. An appointment was made with the mortgage brokers, John Charcol. Details of the property, i.e. the initial cost, the projected expenditure and the anticipated rent, were prepared together with data on current income, house and mortgage commitments.

Alternative mortgages were offered on the basis of remortgaging the clients' existing home. The total fees for arranging this would have been around £2000. Term insurance to cover the loan was required in addition to monthly repayments of around £700 per month. John Charcol's fee would have been reduced if life cover was taken. Interest rates were very competitive at the time and averaged 5 per cent. These offers seemed well worth considering; however, the developers were advised by their solicitor that remortgaging would be as complex a process as the initial house purchase and John Charcol suggested that another approach should be made to the

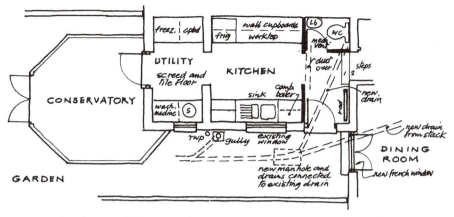

Figure 11.10   The ground floor proposals

developers' existing building society, with the projected rent and resale figures.

The developers approached the commercial department of their society and were then offered a loan of £80 000 for the project at a rate of 1 per cent over the society's standard rate, less an administration fee of £800. It was decided to proceed with the project on this basis.

### Letting the house

The rent finally recommended by the agent was £735 pcm, lower than the figure originally suggested, partly because the property was not ready for occupation until the autumn rather than the summer and perhaps because at the time of the first discussion

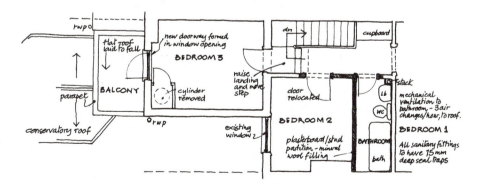

Figure 11.11   The first floor proposals

the agent might not have been aware of the rather shabby state of the street. Nevertheless, it was let very quickly and satisfactorily but only for the minimum period of six months. The house was let unfurnished on an assured shorthold tenancy. A further tenancy of the same length followed at a slightly higher rent. However, it was clear that the occupants had difficulty finding the rent, which became a cause for anxiety.

Both tenancies had created trouble with neighbours over noise, and conversely noise from the adjoining properties seemed likely to be a problem in the future. Children scampering up and down staircases in terraced houses can be quite disturbing and loud music and arguments even more so.

The time when the house would be required for the owners' home had receded, so it was decided to take advantage of an upturn in the market and put the house up for sale.

### Selling the house

The owners asked the estate agents for an indication of a suitable asking price and whether it would be possible to show people the property with the tenants in occupation. Evidently the latter was not normally a problem and an asking price of £115 000 was agreed. The market seemed positive at last. But no offers were made.

Eventually the estate agents reported that the problem was the state of the house as lived in by the tenants. They added that the front of the adjoining house did not help. The owners visited and were rather shocked to find that contrary to the terms of the lease, two rooms had been painted, one quite well and the other very patchily in garish colours. A lot of heavy furniture had been installed, making the small rooms look smaller, and the general effect was displeasing. So the house was temporarily withdrawn from the market.

After the tenants left in mid-November 1996, the house was thoroughly cleaned and partly redecorated. Rugs, mirrors and curtains were reinstated and the house was put back on the market. An added bonus was that the shabby front of the adjoining house had been painted in the interim.

Because it was now past the best time of the year for selling and interest had already been adversely affected by its previous state, the house was put back on the market for the lower price of £110 000. A sale was agreed before Christmas and was completed in mid-February 1997.

## Evaluation of the project

In several ways the project could be considered a limited success, firstly because it had helped to clarify the owners' requirements for a suitable house for retirement, secondly because it made a moderate profit and thirdly because at least some of the occupants had found it a pleasant place to live. However, in other ways it fell between two stools, because it was not the ideal solution for letting or for selling as a development.

According to letting agents, premises suitable for letting should ideally be small because this limits the number of people likely to need to share. Sharing may be restricted by the lease, but unless a landlord lives next door no one knows what happens.

Small premises also usually cost less, so less rent is needed as a return on capital, and if problems arise they are proportionally smaller. Two small units with leases changing at different times are less likely to be a problem than one large one.

A design for selling would have concentrated on character and quality; for example, items such as restoring the windows and fireplaces, a luxurious shower rather than a small bathroom, maybe at the expense of the conservatory. Improving the concrete paving at the front and redesigning the garden patio at the rear would have made the house feel more expensive and attractive, and for the same outlay but less work would probably have attracted more interest, and realized the initial asking price in less time.

## Pointer for subsequent projects

- It is important to keep resale requirements as the top consideration, because circumstances change unexpectedly.
- It is better not to try to sell a house with tenants in situ; they have no interest in being helpful and may put off potential purchasers. Also, some solicitors will not allow their clients to exchange contracts until the occupiers have left, so there would be no time advantage to be gained.

331

# RIBA Plan of Work

Whatever the circumstances, it is vital for the architect to analyse the special requirements of each project to identify any variation in the methods of working set out in the Plan of Work and the Architect's Job Book. Flexibility in management will be essential to handle such variations. Fundamental actions must be taken in good time to implement the agreed procurement path.

## RIBA Plan of Work stages

These outline descriptions recognize developments in the methods of working which have evolved since the original Plan of Work was published in the 1960s. Their essential contents are unchanged. The precise operations in each stage may be modified when carried out in different sequences and conditions.

### A  Inception

Client establishes basic requirements, cost ranges, timetables etc. He appoints architect and principal consultants. Basic project organization is established.

### B  Feasibility

The design team is organized. The brief is developed as fully as possible. The site, legal and other constraints are studied. Alternative design options are considered. The client is advised about the feasibility of the project in functional, technical, financial and contractual terms. His decision is sought on how the project is to proceed.

## C  Outline proposals

The brief is further developed in line with the general approach to layout, design, construction and services. A cost plan is established. The client is asked for his authoritative approval on how to proceed.

## D  Scheme design

The brief is completed and architectural, engineering and services designs are integrated. The cost plan, overall programme and outline specification are developed and planning and other approvals applied for. A report is submitted to the client for his approval.

## E  Detail design

The team designs, co-ordinates and specifies all parts and components, completes cost checks and obtains client's approval of significant details and costs. Specialist tenders may be sought.

## F  Production information

The team prepares working drawings, schedules and specifications and agrees with the client how the work is to be carried out. Specialist tenders may be sought.

## G  Bills of quantities

Bills of quantities are prepared and all documents and arrangements for obtaining tenders are completed. Specialist tenders may be sought.

## H  Tender action

Main contract tenders are obtained by negotiation or competitive tendering procedures. The client is asked to agree that suitable tenders are accepted.

## J  Project planning

Contract documents are processed. The contractor receives information needed to plan the work. The site inspectorate is briefed and all roles are defined. The site is made available for work to start.

## K  Operations on site

Contract is administered and contractual obligations fulfilled with progress and quality control monitored. Financial control, with regular reports to the client, is maintained.

## L  Completion

Project is handed over for occupation. Defects are corrected, claims are resolved and final account is agreed. Final Certificate is issued.

## M  Feedback

The performance of the building and the design and construction teams is analysed and recorded for future reference.

# The RICS/ISVA Homebuyer Survey and Valuation

## Section A    Introduction

- Object
- Action
- Overall opinion

## Section B    The property and location

### B1    The property
- Type and age
- Construction
- Accommodation
- Garage and grounds

### B2    The location

### B3    Circumstances of the inspection

## Section C    The building

### C1    Movement

### C2    Timber defects

### C3    Dampness
- Damp-proof course
- Rising and penetrating damp
- Condensation

## C4   Insulation
## C5   The exterior

- Roof structure and covering
- Chimneys
- Rain water disposal
- Main walls
- Windows etc.
- External decoration
- Other

## C6   The interior

- Loft space
- Floors
- Ceilings
- Internal walls and partitions
- Fireplaces etc.
- Internal joinery
- Internal decoration
- Other

# Section D   The services and site

## D1   Services

- Electricity
- Gas
- Water
- Heating
- Other

## D2   Drainage
## D3   The site

- Garage and outbuildings
- Grounds and boundaries

## Section E   Legal and other matters

*E1   Tenure*

*E2   Regulations etc.*

*E3   Insurance and guarantees*

*E4   Other matters*

## Section F   Summary

*F1   Action*
- Further investigations
- Urgent repairs

*F2   Maintenance considerations*

*F3   Other considerations*

## Section G   Valuation

*G1   Open market value*

*G2   Insurance cover (Reinstatement cost)*

# Association of Corporate Approved Inspectors

## List of members: as at 13 October 1997

### Companies

BAA Building Control Services Ltd
Norfolk House
South Terminal
Gatwick RH6 0JN
Tel: 01293 504860

NHBC Building Control
Services Ltd
Buildmark House
Chiltern Avenue
Amersham HP6 5AP
Tel: 01494 431857

Maunsell Associates
390 Kenton Road
Harrow HA3 9DS
Tel: 0181 909 1891

Butler & Young Associates
Building Control and Fire
Safety Consultants Ltd
433 London Road
Croydon CR0 3PF
Tel: 0181 680 1500

BRCS (Building Control) Ltd
31 Worship Street
London EC2A 2DX
Tel: 0171 588 1100

TPS Special Services Ltd
The Landsdowne Building
2 Landsdowne Road
Croydon CRO 2BX
Tel: 0181 236 4188

## Individuals

Terry R. Atkinson
Timans Associates
West Midlands House
Gipsy Lane
Willenhall WV13 2HA
Tel: 01902 482436

Frank Claybrooke
Building Engineer & Surveyor
79 Dunvant Road
Killay
Swansea SA2 7NL
Tel: 01792 208486
Glynne Holmes
Glynne Holmes Surveying
22 Clos Lancaster
Llantrissant
Pontyclun CF72 8QP
Tel: 01443 225678

Peter Gass
Building Design Consultant
58 Thorntons Close
Pelton
Chester-le-Street DH2 1QH
Tel: 0191 370 0728

Anthony J. Ley
Chartered Building Surveyor & Engineer
'Foreay'
Lower Park Road
Braunton EX33 2HG
Tel: 01271 814395

Martin J. Lowe
AMA Design, Architects & Surveyors
'Westlands'
45 Merton Road
Bootle
Liverpool L20 7AP
Tel: 0151 922 0111

Roger Mahoney
Gerry Lytle Associates Architects
The Fountain Head
Quarry Street
Guildford GU1 3UY
Tel: 01483 301661

John Miller
John Miller Partnership
8 Grassy Glade
Gillingham ME7 3RR
Tel: 01634 233154

Thomas M. Sammons
Chartered Surveyor & Engineer
20 Belsize Close
St Albans AL4 9YD
Tel: 01727 840723

John Spencer
John Spencer Associates
'Roseneath'
Church Lane
Brightwell-cum-Sotwell
Wallingford OX10 0SD
Tel: 01491 839193

# Forms for house purchase

*(from Oyez, London or Birmingham)*

1  Draft Contract: incorporating the National Conditions of Sale. Sent by vendor/solicitor. Own form or OYEZ standard. SCS1

2  Enquiries before Contract: to be answered by vendor/solicitor. Form CON29 (long)

3  HM Land Registry Office Copy of Register – gives title (absolute etc.), easement, covenants and registration no. of property. Sent by vendor/solicitor with entry at last purchase.

4  Application for Office Copies to Register – to Land Registry for any new entries. Form 109

5  Official search with priority of the land – to Land Registry to reserve time for registration. Form 94A

6  Searches from Local Authority, i.e. council interest. Matters affecting property such as road schemes etc. LLC1/CON 29A or D

7  Search for Bankruptcy Only – to Land Charges at Plymouth. K16

8  Official Certificate of Result of Search – from Land Registry giving date priority expires. Form 94D

9  Discharge of Registered Charge – required from vendors if they have a mortgage – to go to Land Registry finally. Form 53

10 Contract (final agreement) – gives title, parties to agreement, purchase price and list of fixtures and fittings. SCS1

11 Transfer document – to go to Land Registry with names of new owners and amount paid. Form 19 (JP)

12 Inland Revenue forms for Stamp Duty – to go to Inland Revenue with payment. Stamps L(A) 451

13 Inland Revenue form for impression – does not appear to be important. Stamps 61

14 Application for new Registration of Title – to go to Land Registry + fee + Transfer Form 19 and Stamps L(A) 451 (+ 53 if mortgaged before).

15 Land Certificate – from Land Registry with acknowledgement of completion of registration. C2A (COMP)

# Specimen contract

1 [The contractor's name] will carry out and complete the work outlined in the attached specification and drawings in a good and workmanlike manner, in accordance with all relevant British standards and codes of practice, to the reasonable satisfaction of [the owner], all for the sum of [the agreed price], plus VAT [if applicable] at the standard rate.

2 The contractor will provide all the materials, equipment, plant and labour necessary to carry out and complete the work.

3 The contractor will begin work on [the starting date], will proceed regularly with the work, and will complete it by [the finishing date], subject only to any changes agreed according to Clause 6. Time is of the essence with regard to this work.

4 Should the contractor fail to finish the work on time without good reason, he agrees to pay the owner damages, which represent the actual loss to the owner of [an agreed figure] for every week or part of a week during which completion is delayed.

5 The contractor will, within 14 days of completing the work, remove all tools, surplus materials and rubbish from the site, leaving it in a clean and tidy condition.

6 Any variation to the work, together with its cost and its effect on the original completion date, will be agreed in writing by both parties before the variation is carried out. Otherwise the completion date will be postponed only if the contractor is prevented from completing the job by factors outside his control.

7　The contractor will comply with all statutory requirements, local and national regulations and by-laws that relate to the work, and will be responsible for making all the required notifications and arranging all necessary site inspections.

8　The contractor will take out appropriate employer's liability insurance and third party liability insurance to cover the work.

9　The contractor will make good at his own expense any damage to the owner's premises caused by him, his employees or his sub-contractors.

10　If the contractor's work is not of a reasonable standard, or if the contractor leaves the site without reasonable explanation for more than [an agreed number] consecutive days, the owner may terminate the contract, paying only for the value of the work done, less compensation for inconvenience or any additional expenses incurred as a result.

11　The contractor will promptly, and at his expense, make good any defects resulting from materials or workmanship which are not in accordance with the terms of the contract.

12　The owner will pay to the contractor [an agreed figure, usually 90 or 95 per cent] of the sum mentioned in Clause 1, or any other such sum as may be agreed in accordance with Clause 6, on submission of the contractor's final account following satisfactory completion of the work. The balance will be paid [an agreed time] from the date of completion, or when all defects arising have been made good in accordance with Clause 11, whichever is the later.

(12a – an alternative to Clause 12 for long projects.)
The owner will make interim payments of [the agreed amounts] to the contractor on satisfactory completion of the following stages of the work [list the stages].

13　The owner and the contractor agree that, should any dispute between them arise, either party will give the other written notice of the dispute. If this cannot be resolved by discussion and negotiation, they will refer the matter to an arbitrator agreed by both parties, whose decision will be final and binding.

(Source: Lawrence, M. (1996) *The Which? Book of Home Improvements*. Which? Ltd.)

# The Housing Grants, Construction and Regeneration Act 1998

Commonly known as the Construction Act, this becomes law on 1 May 1998. There are two parts which are of particular interest to residential developers and householders. These are briefly described here:

## Part I  Grants etc. for renewal of private sector housing

This changes the arrangements for grants aiding domestic improvement works. Except in the case of disabled facilities, grants are now all discretionary, i.e. not mandatory. Quite simply this means that some local authorities will not have sufficient funds to award grants. There are four types:

- *Renovation grants* – for improvements or repairs. These are not payable for homes or conversions less than 10 years old unless for houses in multiple occupation or for the disabled. They can be given to owners or tenants but not for conversions in the latter case and only if the work is required as a term of the tenancy. The grant requires the applicant to have lived in the dwelling for at least 3 years. The purposes for which these grants are given are as follows:
  - unfit premises
  - substantial repair
  - thermal insulation
  - space heating

- satisfactory internal arrangements
- means of escape
- provision of services

● *Common parts grants* – specifically for the improvement of blocks of flats in which there must be 'occupying tenants'. Applicants should be the landlord or 75 per cent of the tenants who as terms of their tenancies are required to improve common parts (lobbies, staircases, lifts and so on). The grants are given for purposes similar to those for renovation grants above.

● *Disabled facilities grants* – can be for landlord or tenant and are mandatory. The main purposes are to improve or provide facilities such as:
  - access inside and outside the dwelling
  - safety generally
  - bathroom facilities – WC, shower/bath, basin etc.
  - kitchen facilities – cooking, washing up, storage etc.
  - heating installation and controls
  - electrical power and lighting installations and fittings

● *Houses in multiple occupation (HMO) grants* – these are for the improvement and repair of existing HMOs or for the provision of new ones by conversion. Only house owners can apply for them. The purposes are similar to those for renovation grants.

● *Home repairs assistance* – this replaces the 'minor works' grants available under the Local Government and Housing Act of 1989. It is an alternative to the four main grants mentioned above and intended specifically for people who are on income support or other benefits. Interestingly it applies also to houseboats and mobile homes.

Except in the case of landlords applying for Common Parts or HMO grants, all applicants will be means tested. The Secretary of State will specify the maximum amounts of grants payable in due course. The local authority must notify an applicant within six months of the application and the work must be done within twelve months of an approval and carried out by the contractor who provides the original estimate. The local authority may have an approved list of contractors. The grant will be paid either in instalments or at the end of the job. If an owner sells the property within five years of receiving a grant, it will be repayable.

### Group repair schemes

The Act sets out the arrangements for these. This is 'enveloping' which used to be carried out in Housing Action Areas or General Improvement Areas. Under the new act it can be carried out anywhere. The grants cover repairs and exterior improvements including those of a structural nature and foundations, in fact 'any part of buildings exposed to the air'. Participants, who should be house owners, are required to contribute to the cost.

## Part II Construction contracts

This part does not apply to a residential occupier but it does apply to written contracts or agreements between developers and others for residential developments. In any case some of the provisions are likely to have repercussions on all future forms of contract so they are briefly described here.

Construction contracts are now deemed to include design or survey-ing work, even 'providing advice on buildings provided it is in writing'.

*Adjudication* – either party to a contract can opt (it is not mandatory) to refer a dispute to an adjudicator. The party must give notice and the adjudicator is to be appointed within the very short time of seven days. The adjudicator is required to reach a decision within twenty-eight days but this can be extended by fourteen days if necessary. At the time of writing there is some debate as to the most suitable body to act as an adjudicator but whoever it is, he or she will have to act quickly. The procedures appear to be very similar to the 'conciliation' procedures featured in engineering contracts and are to be welcomed as an alternative to wasteful arbitration procedures which are universally disliked.

*Payments* – the Act tightens up contract procedures with respect to methods of payment which have become increasingly unsatisfactory in recent years. Parties are to agree 'dates on which sums become due' and the 'final date of payment' and a party may give notice of payment needed not later than five days after the due date. If money is withheld for any reason, notice must be given stating the reason for it and parties must agree how long before the final date for payment such a notice should be served. If payment is not made in full by the final date and no notice to withhold has been served, a party has the right to suspend performance but seven days notice must be given. No longer will it be acceptable for a contractor to withhold payment to a sub-contractor, for example, because the employer has not paid the contractor (the notorious 'pay when paid' practice).

As an aside, it is interesting to note that under the 'Late Payment of Commercial Debts (Interest) Bill' due in late 1998, small firms will be able to charge larger businesses that fail to pay on time 8 per cent interest for up to six years after the dispute arises. Some argue that 8 per cent is not enough and that unauthorized overdraft rates should apply!

*The Scheme for Construction Contracts* – to be published as a Statutory Instrument. It will set out in contract terms the provisions for adjudication and payment procedures and is likely to have a significant effect on standard forms of contract. Those in use at the time of writing in 1998 will be amended in the near future. The effect of this part of the Construction Act on practice has yet to be fully assessed but the fact that the provisions will apply to most construction contracts, including professional service contracts with architects, engineers and surveyors, suggests that few will remain untouched by them.

# Index